Modern Optical Engineering

Modern Optical Engineering

Edited by
Gavin Moore

Larsen & Keller
www.larsen-keller.com

Modern Optical Engineering
Edited by Gavin Moore
ISBN: 978-1-63549-207-1 (Hardback)

© 2017 Larsen & Keller

▤ Larsen & Keller

Published by Larsen and Keller Education,
5 Penn Plaza,
19th Floor,
New York, NY 10001, USA

Cataloging-in-Publication Data

Modern optical engineering / edited by Gavin Moore.
    p. cm.
Includes bibliographical references and index.
ISBN 978-1-63549-207-1
1. Optical engineering. 2. Optics. 3. Optical instruments.
I. Moore, Gavin.
TA1520 .M63 2017
621.36--dc23

The publisher's policy is to use permanent paper from mills that operate a sustainable forestry policy. Furthermore, the publisher ensures that the text paper and cover boards used have met acceptable environmental accreditation standards.

Printed and bound in the United States of America.

For more information regarding Larsen and Keller Education and its products, please visit the publisher's website www.larsen-keller.com

Table of Contents

Permissions

Index

Preface

This book elucidates new techniques and their applications of optical engineering in a multidisciplinary approach. It provides comprehensive insights into this field. Optical engineering refers to the study of optics and designing of optical instruments like microscopes, telescopes, optic discs, lasers, and fiber optic communication systems, etc. This text is a compilation of chapters that discuss the most vital concepts in the field of optical engineering. It presents this complex subject in the most comprehensible and easy to understand language. Also included in it is a detailed explanation of the various concepts and applications of optical engineering. This textbook is a complete source of knowledge on the present status of this important field and it is meant for students who are looking for an elaborate text on modern optical engineering.

To facilitate a deeper understanding of the contents of this book a short introduction of every chapter is written below:

Chapter 1- Optical engineering focuses on optical technologies, such as lenses and telescopes. The aim of optical engineering is to build devices that makes use of the properties of light. This chapter is an overview of the subject matter incorporating all the major aspects of optical engineering.

Chapter 2- The major concepts of optics are discussed in this chapter. It focuses on concepts such as reflection, refraction, diffraction, scattering etc. which are crucial for understanding optical devices. The following text elucidates the crucial concepts of optics, and strategically incorporates all the essential components of optics.

Chapter 3- This chapter will provide an integrated understanding of the design and analytical aspects of optical instrument. This chapter educates the reader on concepts as lens, prism, microscope, telescope, photometer, polarimeter, lensmeter, waveplate etc. The chapter strategically encompasses and incorporates the major components and key concepts of optical engineering, providing a complete understanding.

Chapter 4- Communication that takes places at a distance by using light to carry information is optical communication while the method that is required to transmit information by light through an optical fiber is fiber optic communication. The aspects elucidated in this chapter are of vital importance for the progress of this field and its allied branches.

Finally, I would like to thank the entire team involved in the inception of this book for their valuable time and contribution. This book would not have been possible without their efforts. I would also like to thank my friends and family for their constant support.

Editor

Introduction to Optical Engineering

Optical engineering focuses on optical technologies, such as lenses and telescopes. The aim of optical engineering is to build devices that makes use of the properties of light. This chapter is an overview of the subject matter incorporating all the major aspects of optical engineering.

Optical Engineering

Optical engineering is the field of study that focuses on applications of optics. Optical engineers design components of optical instruments such as lenses, microscopes, telescopes, and other equipment that utilizes the properties of light. Other devices include optical sensors and measurement systems, lasers, fiber optic communication systems, optical disc systems (e.g. CD, DVD), etc.

Because optical engineers want to design and build devices that make light do something useful, they must understand and apply the science of optics in substantial detail, in order to know what is physically possible to achieve (physics and chemistry). However, they also must know what is practical in terms of available technology, materials, costs, design methods, etc. As with other fields of engineering, computers are important to many (perhaps most) optical engineers. They are used with instruments, for simulation, in design, and for many other applications. Engineers often use general computer tools such as spreadsheets and programming languages, and they make frequent use of specialized optical software designed specifically for their field.

Optical engineering metrology uses optical methods to measure micro-vibrations with instruments like the laser speckle interferometer or to measure the properties of the various masses with instruments measuring refraction.

Optics

Optics is the branch of physics which involves the behaviour and properties of light, including its interactions with matter and the construction of instruments that use or detect it. Optics usually describes the behaviour of visible, ultraviolet, and infrared light. Because light is an electromagnetic wave, other forms of electromagnetic radiation such as X-rays, microwaves, and radio waves exhibit similar properties.

Most optical phenomena can be accounted for using the classical electromagnetic description of light. Complete electromagnetic descriptions of light are, however, often difficult to apply in practice. Practical optics is usually done using simplified models. The most common of these, geometric optics, treats light as a collection of rays that travel in straight lines and bend when they pass through or reflect from surfaces. Physical optics is a more comprehensive model of light, which includes wave effects such as diffraction and interference that cannot be accounted for in geometric optics. Historically, the ray-based model of light was developed first, followed by the wave model of light. Progress in electromagnetic theory in the 19th century led to the discovery that light waves were in fact electromagnetic radiation.

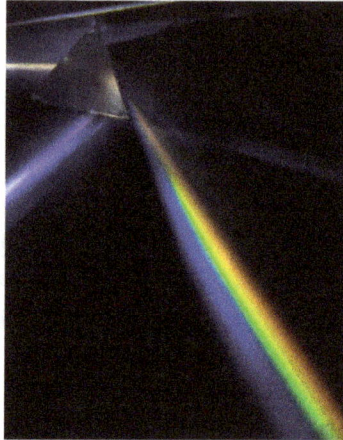

Optics includes study of dispersion of light.

Some phenomena depend on the fact that light has both wave-like and particle-like properties. Explanation of these effects requires quantum mechanics. When considering light's particle-like properties, the light is modelled as a collection of particles called "photons". Quantum optics deals with the application of quantum mechanics to optical systems.

Optical science is relevant to and studied in many related disciplines including astronomy, various engineering fields, photography, and medicine (particularly ophthalmology and optometry). Practical applications of optics are found in a variety of technologies and everyday objects, including mirrors, lenses, telescopes, microscopes, lasers, and fibre optics.

History

Optics began with the development of lenses by the ancient Egyptians and Mesopotamians. The earliest known lenses, made from polished crystal, often quartz, date from as early as 700 BC for Assyrian lenses such as the Layard/Nimrud lens. The ancient Romans and Greeks filled glass spheres with water to make lenses. These practical developments were followed by the development of theories of light and vision by ancient Greek and Indian philosophers, and the development of geometrical optics in the Greco-Roman world.

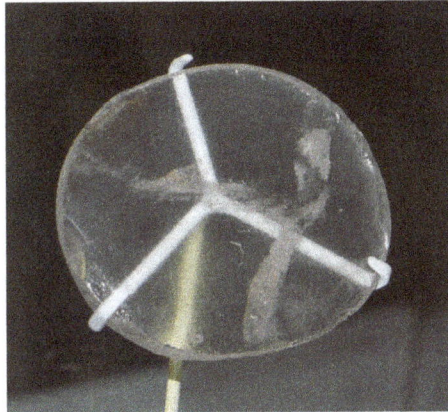

The Nimrud lens

Greek philosophy on optics broke down into two opposing theories on how vision worked, the "intromission theory" and the "emission theory". The intro-mission approach saw vision as coming from objects casting off copies of themselves (called eidola) that were captured by the eye. With many propagators including Democritus, Epicurus, Aristotle and their followers, this theory seems to have some contact with modern theories of what vision really is, but it remained only speculation lacking any experimental foundation.

Plato first articulated the emission theory, the idea that visual perception is accomplished by rays emitted by the eyes. He also commented on the parity reversal of mirrors in *Timaeus*. Some hundred years later, Euclid wrote a treatise entitled *Optics* where he linked vision to geometry, creating *geometrical optics*. He based his work on Plato's emission theory wherein he described the mathematical rules of perspective and described the effects of refraction qualitatively, although he questioned that a beam of light from the eye could instantaneously light up the stars every time someone blinked. Ptolemy, in his treatise *Optics*, held an extramission-intromission theory of vision: the rays (or flux) from the eye formed a cone, the vertex being within the eye, and the base defining the visual field. The rays were sensitive, and conveyed information back to the observer's intellect about the distance and orientation of surfaces. He summarised much of Euclid and went on to describe a way to measure the angle of refraction, though he failed to notice the empirical relationship between it and the angle of incidence.

Alhazen (Ibn al-Haytham), "the father of Optics"

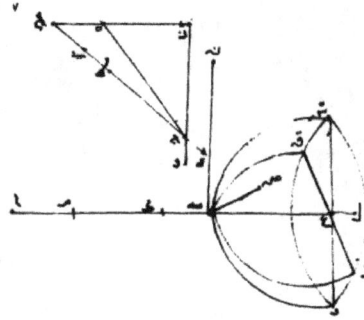

Reproduction of a page of Ibn Sahl's manuscript showing his knowledge of the law of refraction.

During the Middle Ages, Greek ideas about optics were resurrected and extended by writers in the Muslim world. One of the earliest of these was Al-Kindi (c. 801–73) who wrote on the merits of Aristotelian and Euclidean ideas of optics, favouring the emission theory since it could better quantify optical phenomenon. In 984, the Persian mathematician Ibn Sahl wrote the treatise "On burning mirrors and lenses", correctly describing a law of refraction equivalent to Snell's law. He used this law to compute optimum shapes for lenses and curved mirrors. In the early 11th century, Alhazen (Ibn al-Haytham) wrote the *Book of Optics* (*Kitab al-manazir*) in which he explored reflection and refraction and proposed a new system for explaining vision and light based on observation and experiment. He rejected the "emission theory" of Ptolemaic optics with its rays being emitted by the eye, and instead put forward the idea that light reflected in all directions in straight lines from all points of the objects being viewed and then entered the eye, although he was unable to correctly explain how the eye captured the rays. Alhazen's work was largely ignored in the Arabic world but it was anonymously translated into Latin around 1200 A.D. and further summarised and expanded on by the Polish monk Witelo making it a standard text on optics in Europe for the next 400 years.

In the 13th century in medieval Europe, English bishop Robert Grosseteste wrote on a wide range of scientific topics, and discussed light from four different perspectives: an epistemology of light, a metaphysics or cosmogony of light, an etiology or physics of light, and a theology of light, basing it on the works Aristotle and Platonism. Grosseteste's most famous disciple, Roger Bacon, wrote works citing a wide range of recently translated optical and philosophical works, including those of Alhazen, Aristotle, Avicenna, Averroes, Euclid, al-Kindi, Ptolemy, Tideus, and Constantine the African. Bacon was able to use parts of glass spheres as magnifying glasses to demonstrate that light reflects from objects rather than being released from them.

The first wearable eyeglasses were invented in Italy around 1286. This was the start of the optical industry of grinding and polishing lenses for these "spectacles", first in Venice and Florence in the thirteenth century, and later in the spectacle making centres in both the Netherlands and Germany. Spectacle makers created improved types of lenses for the correction of vision based more on empirical knowledge gained from observing the effects of the lenses rather than using the rudimentary optical theory of the day (theory which for the most part could not even adequately explain how spectacles worked). This practical development, mastery, and experimentation with lenses led directly to the invention of the compound optical microscope around 1595, and the refracting telescope in 1608, both of which appeared in the spectacle making centres in the Netherlands.

In the early 17th century Johannes Kepler expanded on geometric optics in his writings, covering lenses, reflection by flat and curved mirrors, the principles of pinhole cameras, inverse-square law governing the intensity of light, and the optical explanations of astronomical phenomena such as lunar and solar eclipses and astronomical parallax. He was also able to correctly deduce the role of the retina as the actual organ that recorded images, finally being able to scientifically quantify the effects of different types of lenses that spectacle makers had been observing over the previous 300 years. After the invention of the telescope Kepler set out the theoretical basis on how they worked and described an improved version, known as the *Keplerian telescope*, using two convex lenses to produce higher magnification.

Optical theory progressed in the mid-17th century with treatises written by philosopher René Descartes, which explained a variety of optical phenomena including reflection and refraction by assuming that light was emitted by objects which produced it. This differed substantively from the ancient Greek emission theory. In the late 1660s and early 1670s, Isaac Newton expanded Descartes' ideas into a corpuscle theory of light, famously determining that white light was a mix of colours which can be separated into its component parts with a prism. In 1690, Christiaan Huygens proposed a wave theory for light based on suggestions that had been made by Robert Hooke in 1664. Hooke himself publicly criticised Newton's theories of light and the feud between the two lasted until Hooke's death. In 1704, Newton published *Opticks* and, at the time, partly because of his success in other areas of physics, he was generally considered to be the victor in the debate over the nature of light.

Newtonian optics was generally accepted until the early 19th century when Thomas Young and Augustin-Jean Fresnel conducted experiments on the interference of light that firmly established light's wave nature. Young's famous double slit experiment showed that light followed the law of superposition, which is a wave-like property not predicted by Newton's corpuscle theory. This work led to a theory of diffraction for light and opened an entire area of study in physical optics. Wave optics was successfully unified with electromagnetic theory by James Clerk Maxwell in the 1860s.

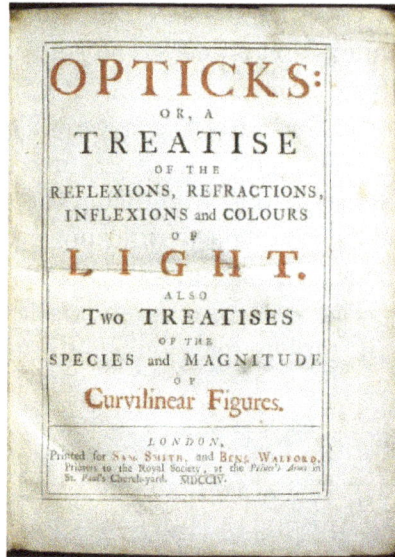

Cover of the first edition of Newton's Opticks

The next development in optical theory came in 1899 when Max Planck correctly modelled blackbody radiation by assuming that the exchange of energy between light and matter only occurred in discrete amounts he called *quanta*. In 1905 Albert Einstein published the theory of the photoelectric effect that firmly established the quantization of light itself. In 1913 Niels Bohr showed that atoms could only emit discrete amounts of energy, thus explaining the discrete lines seen in emission and absorption spectra. The understanding of the interaction between light and matter which followed from these developments not only formed the basis of quantum optics but also was crucial for the development of quantum mechanics as a whole. The ultimate culmination, the theory of quantum electrodynamics, explains all optics and electromagnetic processes in general as the result of the exchange of real and virtual photons.

Quantum optics gained practical importance with the inventions of the maser in 1953 and of the laser in 1960. Following the work of Paul Dirac in quantum field theory, George Sudarshan, Roy J. Glauber, and Leonard Mandel applied quantum theory to the electromagnetic field in the 1950s and 1960s to gain a more detailed understanding of photodetection and the statistics of light.

Classical Optics

Classical optics is divided into two main branches: geometrical (or ray) optics and physical (or wave) optics. In geometrical optics, light is considered to travel in straight lines, while in physical optics, light is considered as an electromagnetic wave.

Geometrical optics can be viewed as an approximation of physical optics that applies when the wavelength of the light used is much smaller than the size of the optical elements in the system being modelled.

Geometrical Optics

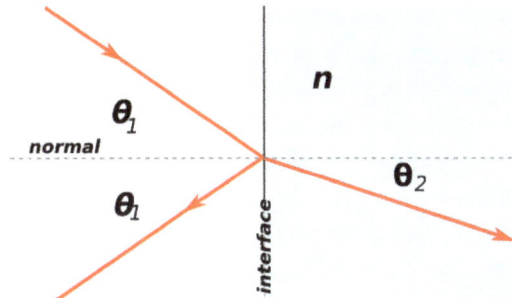

Geometry of reflection and refraction of light rays

Geometrical optics, or *ray optics*, describes the propagation of light in terms of "rays" which travel in straight lines, and whose paths are governed by the laws of reflection and refraction at interfaces between different media. These laws were discovered empirically as far back as 984 AD and have been used in the design of optical components and instruments from then until the present day. They can be summarised as follows:

When a ray of light hits the boundary between two transparent materials, it is divided into a reflected and a refracted ray.

> The law of reflection says that the reflected ray lies in the plane of incidence, and the angle of reflection equals the angle of incidence.

> The law of refraction says that the refracted ray lies in the plane of incidence, and the sine of the angle of refraction divided by the sine of the angle of incidence is a constant:

$$\frac{\sin \theta_1}{\sin \theta_2} = n,$$

where n is a constant for any two materials and a given colour of light. If the first material is air or vacuum, n is the refractive index of the second material.

The laws of reflection and refraction can be derived from Fermat's principle which states that *the path taken between two points by a ray of light is the path that can be traversed in the least time.*

Approximations

Geometric optics is often simplified by making the paraxial approximation, or "small angle approximation". The mathematical behaviour then becomes linear, allowing optical components and systems to be described by simple matrices. This leads to the techniques of Gaussian optics and *paraxial ray tracing*, which are used to find basic properties of optical systems, such as approximate image and object positions and magnifications.

Reflections

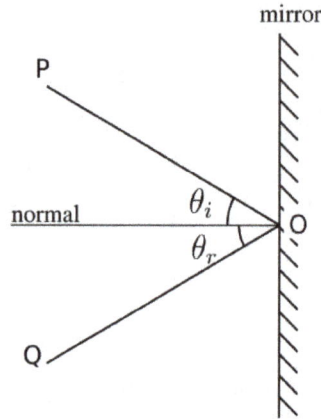

Diagram of specular reflection

Reflections can be divided into two types: specular reflection and diffuse reflection. Specular reflection describes the gloss of surfaces such as mirrors, which reflect light in a simple, predictable way. This allows for production of reflected images that can be associated with an actual (real) or extrapolated (virtual) location in space. Diffuse reflection describes non-glossy materials, such as paper or rock. The reflections from these surfaces can only be described statistically, with the exact distribution of the reflected light depending on the microscopic structure of the material. Many diffuse reflectors are described or can be approximated by Lambert's cosine law, which describes surfaces that have equal luminance when viewed from any angle. Glossy surfaces can give both specular and diffuse reflection.

In specular reflection, the direction of the reflected ray is determined by the angle the incident ray makes with the surface normal, a line perpendicular to the surface at the point where the ray hits. The incident and reflected rays and the normal lie in a single plane, and the angle between the reflected ray and the surface normal is the same as that between the incident ray and the normal. This is known as the Law of Reflection.

For flat mirrors, the law of reflection implies that images of objects are upright and the same distance behind the mirror as the objects are in front of the mirror. The image size is the same as the object size. The law also implies that mirror images are parity inverted, which we perceive as a left-right inversion. Images formed from reflection in two (or any even number of) mirrors are not parity inverted. Corner reflectors retroreflect light, producing reflected rays that travel back in the direction from which the incident rays came.

Mirrors with curved surfaces can be modelled by ray-tracing and using the law of reflection at each point on the surface. For mirrors with parabolic surfaces, parallel rays incident on the mirror produce reflected rays that converge at a common focus. Other curved surfaces may also focus light, but with aberrations due to the diverging shape causing the focus to be smeared out in space. In particular, spherical mirrors exhibit spherical aberration. Curved mirrors can form images with magnification greater than or less than one, and the magnification can be negative, indicating that the image is inverted. An upright image formed by reflection in a mirror is always virtual, while an inverted image is real and can be projected onto a screen.

Refractions

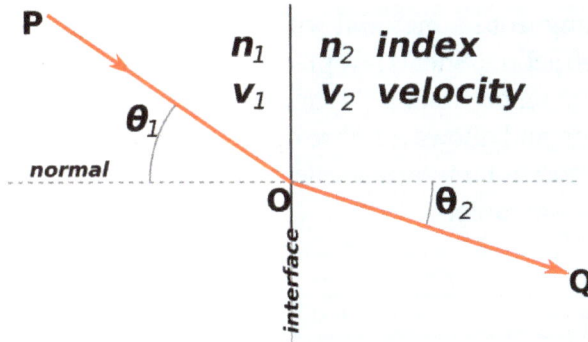

P

n_1 | n_2 *index*
v_1 | v_2 *velocity*

θ_1

normal

O

interface

θ_2

Q

Illustration of Snell's Law for the case $n_1 < n_2$, such as air/water interface

Refraction occurs when light travels through an area of space that has a changing index of refraction; this principle allows for lenses and the focusing of light. The simplest case of refraction occurs when there is an interface between a uniform medium with index of refraction n_1 and another medium with index of refraction n_2. In such situations, Snell's Law describes the resulting deflection of the light ray:

$$n_1 \sin \theta_1 = n_2 \sin \theta_2$$

where θ_1 and θ_2 are the angles between the normal (to the interface) and the incident and refracted waves, respectively.

The index of refraction of a medium is related to the speed, v, of light in that medium by

$$n = c / v ,$$

where c is the speed of light in vacuum.

Snell's Law can be used to predict the deflection of light rays as they pass through linear media as long as the indexes of refraction and the geometry of the media are known. For example, the propagation of light through a prism results in the light ray being deflected depending on the shape and orientation of the prism. In most materials, the index of refraction varies with the frequency of the light. Taking this into account, Snell's Law can be used to predict how a prism will disperse light into a spectrum. The discovery of this phenomenon when passing light through a prism is famously attributed to Isaac Newton.

Some media have an index of refraction which varies gradually with position and, thus, light rays in the medium are curved. This effect is responsible for mirages seen on hot days: a change in index of refraction air with height causes light rays to bend, creating the appearance of specular reflections in the distance (as if on the surface of a pool of water). Optical materials with varying index of refraction are called gradient-index (GRIN) materials. Such materials are used to make gradient-index optics.

For light rays travelling from a material with a high index of refraction to a material with a low index of refraction, Snell's law predicts that there is no θ_2 when θ_1 is large. In this case, no transmission occurs; all the light is reflected. This phenomenon is called total internal reflection and allows for fibre optics technology. As light travels down an optical fibre, it undergoes total internal reflection allowing for essentially no light to be lost over the length of the cable.

Lenses

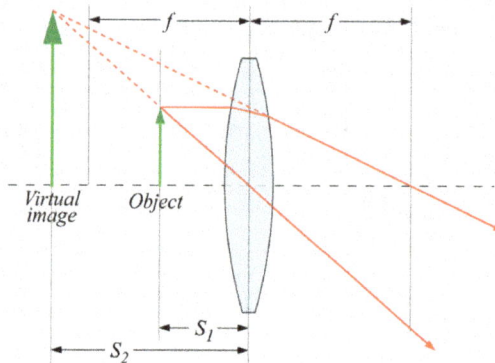

A ray tracing diagram for a converging lens.

A device which produces converging or diverging light rays due to refraction is known as a *lens*. Lenses are characterized by their focal length: a converging lens has positive focal length, while a diverging lens has negative focal length. Smaller focal length indicates that the lens has a stronger converging or diverging effect. The focal length of a simple lens in air is given by the lensmaker's equation.

Ray tracing can be used to show how images are formed by a lens. For a thin lens in air, the location of the image is given by the simple equation

$$\frac{1}{S_1} + \frac{1}{S_2} = \frac{1}{f},$$

where S_1 is the distance from the object to the lens, S_2 is the distance from the lens to the image, and f is the focal length of the lens. In the sign convention used here, the object and image distances are positive if the object and image are on opposite sides of the lens.

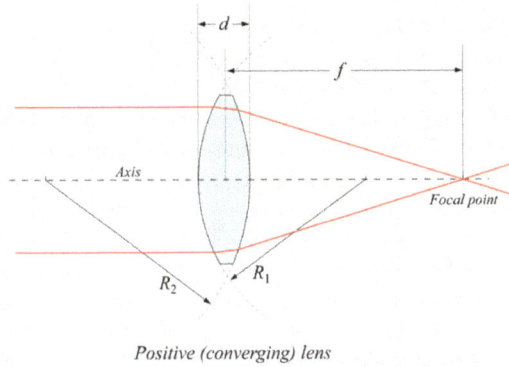

Positive (converging) lens

Incoming parallel rays are focused by a converging lens onto a spot one focal length from the lens, on the far side of the lens. This is called the rear focal point of the lens. Rays from an object at finite distance are focused further from the lens than the focal distance; the closer the object is to the lens, the further the image is from the lens.

With diverging lenses, incoming parallel rays diverge after going through the lens, in such a way that they seem to have originated at a spot one focal length in front of the lens. This is the lens's front focal point. Rays from an object at finite distance are associated with a virtual image that is closer to the lens than the focal point, and on the same side of the lens as the object. The closer the object is to the lens, the closer the virtual image is to the lens. As with mirrors, upright images produced by a single lens are virtual, while inverted images are real.

Lenses suffer from aberrations that distort images. *Monochromatic aberrations* occur because the geometry of the lens does not perfectly direct rays from each object point to a single point on the image, while chromatic aberration occurs because the index of refraction of the lens varies with the wavelength of the light.

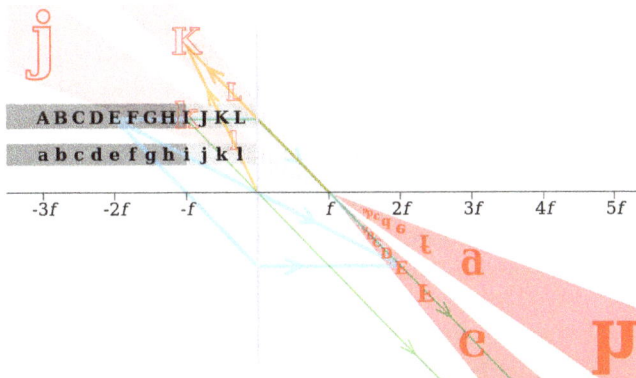

Images of black letters in a thin convex lens of focal length f are shown in red. Selected rays are shown for letters E, I and K in blue, green and orange, respectively. Note that E (at 2f) has anequal-size, real and inverted image; I (at f) has its image at infinity; and K (at f/2) has a double-size, virtual and upright image.

Physical Optics

In physical optics, light is considered to propagate as a wave. This model predicts phenomena such as interference and diffraction, which are not explained by geometric optics. The speed of light waves in air is approximately 3.0×10^8 m/s (exactly 299,792,458 m/s in vacuum). The wavelength of visible light waves varies between 400 and 700 nm, but the term "light" is also often applied to infrared (0.7–300 μm) and ultraviolet radiation (10–400 nm).

The wave model can be used to make predictions about how an optical system will behave without requiring an explanation of what is "waving" in what medium. Until the middle of the 19th century, most physicists believed in an "ethereal" medium in which the light disturbance propagated. The existence of electromagnetic waves was predicted in 1865 by Maxwell's equations. These waves propagate at the speed of light and have varying electric and magnetic fields which are orthogonal to one another, and also to the direction of propagation of the waves. Light waves are now generally treated as electromagnetic waves except when quantum mechanical effects have to be considered.

Modelling and Design of Optical Systems Using Physical Optics

Many simplified approximations are available for analysing and designing optical systems. Most of these use a single scalar quantity to represent the electric field of the light wave, rather than using a vector model with orthogonal electric and magnetic vectors. The Huygens–Fresnel equation is one such model. This was derived empirically by Fresnel in 1815, based on Huygens' hypothesis that each point on a wavefront generates a secondary spherical wavefront, which Fresnel combined with the principle of superposition of waves. The Kirchhoff diffraction equation, which is derived using Maxwell's equations, puts the Huygens-Fresnel equation on a firmer physical foundation. Examples of the application of Huygens–Fresnel principle can be found in the sections on diffraction and Fraunhofer diffraction.

More rigorous models, involving the modelling of both electric and magnetic fields of the light wave, are required when dealing with the detailed interaction of light with materials where the interaction depends on their electric and magnetic properties. For instance, the behaviour of a light wave interacting with a metal surface is quite different from what happens when it interacts with a dielectric material. A vector model must also be used to model polarised light.

Numerical modeling techniques such as the finite element method, the boundary element method and the transmission-line matrix method can be used to model the propagation of light in systems which cannot be solved analytically. Such models are computationally demanding and are normally only used to solve small-scale problems that require accuracy beyond that which can be achieved with analytical solutions.

All of the results from geometrical optics can be recovered using the techniques of Fourier optics which apply many of the same mathematical and analytical techniques used in acoustic engineering and signal processing.

Gaussian beam propagation is a simple paraxial physical optics model for the propagation of coherent radiation such as laser beams. This technique partially accounts for diffraction, allowing accurate calculations of the rate at which a laser beam expands with distance, and the minimum size to which the beam can be focused. Gaussian beam propagation thus bridges the gap between geometric and physical optics.

Superposition and Interference

In the absence of nonlinear effects, the superposition principle can be used to predict the shape of interacting waveforms through the simple addition of the disturbances. This interaction of waves to produce a resulting pattern is generally termed "interference" and can result in a variety of outcomes. If two waves of the same wavelength and frequency are *in phase*, both the wave crests and wave troughs align. This results in constructive interference and an increase in the amplitude of the wave, which for light is associated with a brightening of the waveform in that location. Alternatively, if the two waves of the same wavelength and frequency are out of phase, then the wave crests will align with wave troughs and vice versa. This results in destructive interference and a decrease in the amplitude of the wave, which for light is associated with a dimming of the waveform at that location.

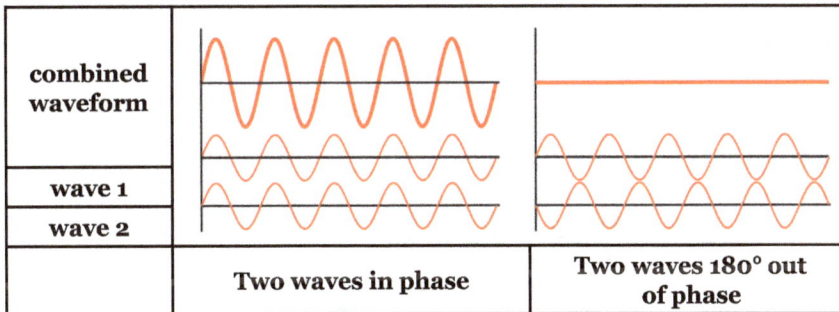

	Two waves in phase	Two waves 180° out of phase
combined waveform		
wave 1		
wave 2		

When oil or fuel is spilled, colourful patterns are formed by thin-film interference.

Since the Huygens–Fresnel principle states that every point of a wavefront is associated with the production of a new disturbance, it is possible for a wavefront to interfere with itself constructively or destructively at different locations producing bright and dark fringes in regular and predictable patterns. Interferometry is the science of measuring these patterns, usually as a means of making precise determinations of distances or angular resolutions. The Michelson interferometer was a famous instrument which used interference effects to accurately measure the speed of light.

The appearance of thin films and coatings is directly affected by interference effects. Antireflective coatings use destructive interference to reduce the reflectivity of the surfaces they coat, and can be used to minimise glare and unwanted reflections. The simplest case is a single layer with thickness one-fourth the wavelength of incident light. The reflected wave from the top of the film and the reflected wave from the film/material interface are then exactly 180° out of phase, causing destructive interference. The waves are only exactly out of phase for one wavelength, which would typically be chosen to be near the centre of the visible spectrum, around 550 nm. More complex designs using multiple layers can achieve low reflectivity over a broad band, or extremely low reflectivity at a single wavelength.

Constructive interference in thin films can create strong reflection of light in a range of wavelengths, which can be narrow or broad depending on the design of the coating. These films are used to make dielectric mirrors, interference filters, heat reflectors, and filters for colour separation in colour television cameras. This interference effect is also what causes the colourful rainbow patterns seen in oil slicks.

Diffraction and Optical Resolution

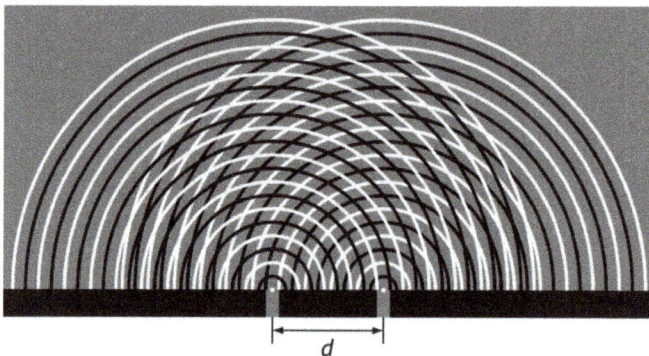

Diffraction on two slits separated by distance d. The bright fringes occur along lines where black lines intersect with black lines and white lines intersect with white lines. These fringes are separated by angle θ and are numbered as order n.

Diffraction is the process by which light interference is most commonly observed. The effect was first described in 1665 by Francesco Maria Grimaldi, who also coined the term from the Latin *diffringere*, 'to break into pieces'. Later that century, Robert Hooke and

Isaac Newton also described phenomena now known to be diffraction in Newton's rings while James Gregory recorded his observations of diffraction patterns from bird feathers.

The first physical optics model of diffraction that relied on the Huygens–Fresnel principle was developed in 1803 by Thomas Young in his interference experiments with the interference patterns of two closely spaced slits. Young showed that his results could only be explained if the two slits acted as two unique sources of waves rather than corpuscles. In 1815 and 1818, Augustin-Jean Fresnel firmly established the mathematics of how wave interference can account for diffraction.

The simplest physical models of diffraction use equations that describe the angular separation of light and dark fringes due to light of a particular wavelength (λ). In general, the equation takes the form

$$m\lambda = d \sin \theta$$

where d is the separation between two wavefront sources (in the case of Young's experiments, it was two slits), θ is the angular separation between the central fringe and the m th order fringe, where the central maximum is $m = 0$.

This equation is modified slightly to take into account a variety of situations such as diffraction through a single gap, diffraction through multiple slits, or diffraction through a diffraction grating that contains a large number of slits at equal spacing. More complicated models of diffraction require working with the mathematics of Fresnel or Fraunhofer diffraction.

X-ray diffraction makes use of the fact that atoms in a crystal have regular spacing at distances that are on the order of one angstrom. To see diffraction patterns, x-rays with similar wavelengths to that spacing are passed through the crystal. Since crystals are three-dimensional objects rather than two-dimensional gratings, the associated diffraction pattern varies in two directions according to Bragg reflection, with the associated bright spots occurring in unique patterns and d being twice the spacing between atoms.

Diffraction effects limit the ability for an optical detector to optically resolve separate light sources. In general, light that is passing through an aperture will experience diffraction and the best images that can be created (as described in diffraction-limited optics) appear as a central spot with surrounding bright rings, separated by dark nulls; this pattern is known as an Airy pattern, and the central bright lobe as an Airy disk. The size of such a disk is given by

$$\sin \theta = 1.22 \frac{\lambda}{D}$$

where θ is the angular resolution, λ is the wavelength of the light, and D is the diameter of the lens aperture. If the angular separation of the two points is significantly less than the Airy disk angular radius, then the two points cannot be resolved in the image, but if their angular separation is much greater than this, distinct images of the two points are formed and they can therefore be resolved. Rayleigh defined the somewhat arbitrary "Rayleigh criterion" that two points whose angular separation is equal to the Airy disk radius (measured to first null, that is, to the first place where no light is seen) can be considered to be resolved. It can be seen that the greater the diameter of the lens or its aperture, the finer the resolution. Interferometry, with its ability to mimic extremely large baseline apertures, allows for the greatest angular resolution possible.

For astronomical imaging, the atmosphere prevents optimal resolution from being achieved in the visible spectrum due to the atmospheric scattering and dispersion which cause stars to twinkle. Astronomers refer to this effect as the quality of astronomical seeing. Techniques known as adaptive optics have been used to eliminate the atmospheric disruption of images and achieve results that approach the diffraction limit.

Dispersion and Scattering

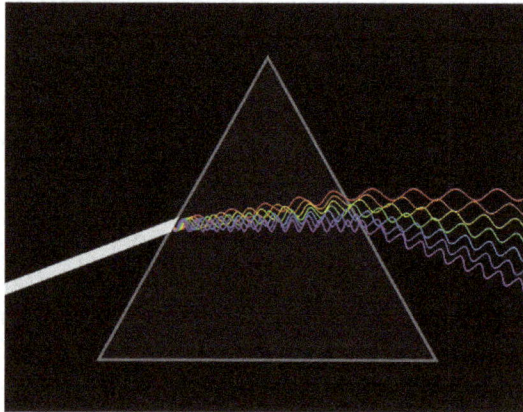

Conceptual animation of light dispersion through a prism. High frequency (blue) light is deflected the most, and low frequency (red) the least.

Refractive processes take place in the physical optics limit, where the wavelength of light is similar to other distances, as a kind of scattering. The simplest type of scattering is Thomson scattering which occurs when electromagnetic waves are deflected by single particles. In the limit of Thomson scattering, in which the wavelike nature of light is evident, light is dispersed independent of the frequency, in contrast to Compton scattering which is frequency-dependent and strictly a quantum mechanical process, involving the nature of light as particles. In a statistical sense, elastic scattering of light by numerous particles much smaller than the wavelength of the light is a process known as Rayleigh scattering while the similar process for scattering by particles that are similar or larger in wavelength is known as Mie scattering with the Tyndall effect being a commonly observed result. A small proportion of light scattering from atoms or molecules may undergo Raman scattering, wherein the frequency changes due to excitation of the atoms and molecules. Brillouin scattering occurs when the frequency of light changes due to local changes with time and movements of a dense material.

Dispersion occurs when different frequencies of light have different phase velocities, due either to material properties (*material dispersion*) or to the geometry of an optical waveguide (*waveguide dispersion*). The most familiar form of dispersion is a decrease in index of refraction with increasing wavelength, which is seen in most transparent materials. This is called "normal dispersion". It occurs in all dielectric materials, in wavelength ranges where the material does not absorb light. In wavelength ranges where a medium has significant absorption, the index of refraction can increase with wavelength. This is called "anomalous dispersion".

The separation of colours by a prism is an example of normal dispersion. At the surfaces of the prism, Snell's law predicts that light incident at an angle θ to the normal will be refracted at an angle arcsin(sin (θ) / n). Thus, blue light, with its higher refractive index, is bent more strongly than red light, resulting in the well-known rainbow pattern.

Dispersion: two sinusoids propagating at different speeds make a moving interference pattern. The red dot moves with the phase velocity, and the green dots propagate with the group velocity. In this case, the phase velocity is twice the group velocity. The red dot overtakes two green dots, when moving from the left to the right of the figure. In effect, the individual waves (which travel with the phase velocity) escape from the wave packet (which travels with the group velocity).

Material dispersion is often characterised by the Abbe number, which gives a simple measure of dispersion based on the index of refraction at three specific wavelengths. Waveguide dispersion is dependent on the propagation constant. Both kinds of dispersion cause changes in the group characteristics of the wave, the features of the wave packet that change with the same frequency as the amplitude of the electromagnetic wave. "Group velocity dispersion" manifests as a spreading-out of the signal "envelope" of the radiation and can be quantified with a group dispersion delay parameter:

$$D = \frac{1}{v_g^2} \frac{dv_g}{d\lambda}$$

where v_g is the group velocity. For a uniform medium, the group velocity is

$$v_g = c\left(n - \lambda \frac{dn}{d\lambda}\right)^{-1}$$

where n is the index of refraction and c is the speed of light in a vacuum. This gives a simpler form for the dispersion delay parameter:

$$D = -\frac{\lambda}{c}\frac{d^2 n}{d\lambda^2}.$$

If D is less than zero, the medium is said to have *positive dispersion* or normal dispersion. If D is greater than zero, the medium has *negative dispersion*. If a light pulse is propagated through a normally dispersive medium, the result is the higher frequency components slow down more than the lower frequency components. The pulse therefore becomes *positively chirped*, or *up-chirped*, increasing in frequency with time. This causes the spectrum coming out of a prism to appear with red light the least refracted and blue/violet light the most refracted. Conversely, if a pulse travels through an anomalously (negatively) dispersive medium, high frequency components travel faster than the lower ones, and the pulse becomes *negatively chirped*, or *down-chirped*, decreasing in frequency with time.

The result of group velocity dispersion, whether negative or positive, is ultimately temporal spreading of the pulse. This makes dispersion management extremely important in optical communications systems based on optical fibres, since if dispersion is too high, a group of pulses representing information will each spread in time and merge, making it impossible to extract the signal.

Polarization

Polarization is a general property of waves that describes the orientation of their oscillations. For transverse waves such as many electromagnetic waves, it describes the orientation of the oscillations in the plane perpendicular to the wave's direction of travel. The oscillations may be oriented in a single direction (linear polarization), or the oscillation direction may rotate as the wave travels (circular or elliptical polarization). Circularly polarised waves can rotate rightward or leftward in the direction of travel, and which of those two rotations is present in a wave is called the wave's chirality.

The typical way to consider polarization is to keep track of the orientation of the electric field vector as the electromagnetic wave propagates. The electric field vector of a plane wave may be arbitrarily divided into two perpendicular components labeled x and y (with z indicating the direction of travel). The shape traced out in the x-y plane by the electric field vector is a Lissajous figure that describes the *polarization state*. The following figures show some examples of the evolution of the electric field vector (blue), with time (the vertical axes), at a particular point in space, along with its x and y components (red/left and green/right), and the path traced by the vector in the plane (purple): The same evolution would occur when looking at the electric field at a particular time while evolving the point in space, along the direction opposite to propagation.

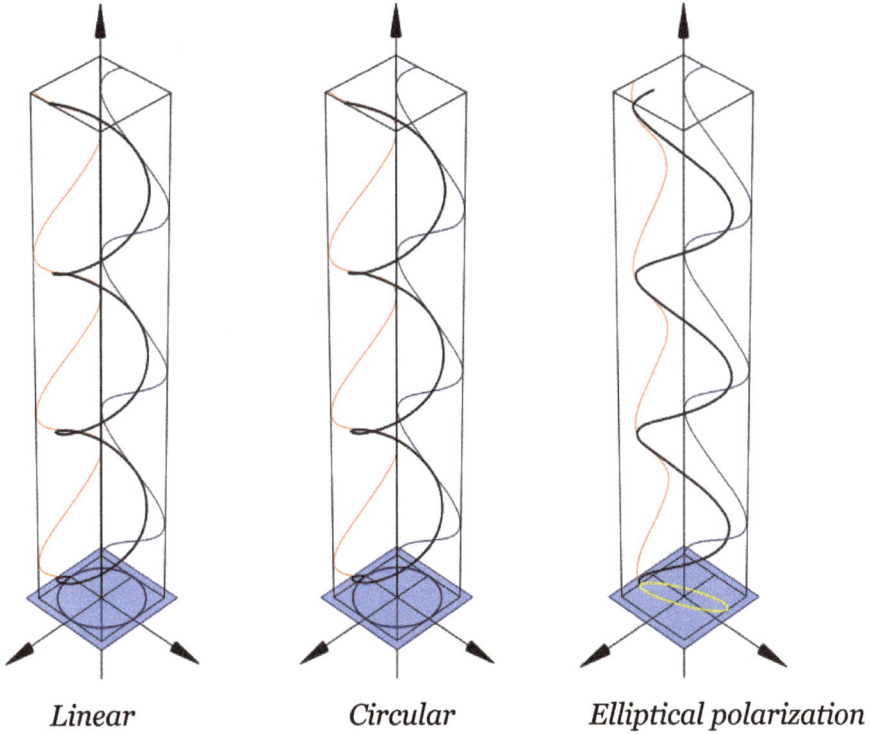

Linear   *Circular*   *Elliptical polarization*

In the leftmost figure above, the x and y components of the light wave are in phase. In this case, the ratio of their strengths is constant, so the direction of the electric vector (the vector sum of these two components) is constant. Since the tip of the vector traces out a single line in the plane, this special case is called linear polarization. The direction of this line depends on the relative amplitudes of the two components.

In the middle figure, the two orthogonal components have the same amplitudes and are 90° out of phase. In this case, one component is zero when the other component is at maximum or minimum amplitude. There are two possible phase relationships that satisfy this requirement: the x component can be 90° ahead of the y component or it can be 90° behind the y component. In this special case, the electric vector traces out a circle in the plane, so this polarization is called circular polarization. The rotation direction in the circle depends on which of the two phase relationships exists and corresponds to *right-hand circular polarization* and *left-hand circular polarization*.

In all other cases, where the two components either do not have the same amplitudes and/or their phase difference is neither zero nor a multiple of 90°, the polarization is called elliptical polarization because the electric vector traces out an ellipse in the plane (the *polarization ellipse*). This is shown in the above figure on the right. Detailed mathematics of polarization is done using Jones calculus and is characterised by the Stokes parameters.

Changing Polarization

Media that have different indexes of refraction for different polarization modes are called *birefringent*. Well known manifestations of this effect appear in optical wave plates/retarders (linear modes) and in Faraday rotation/optical rotation (circular modes). If the path length in the birefringent medium is sufficient, plane waves will exit the material with a significantly different propagation direction, due to refraction. For example, this is the case with macroscopic crystals of calcite, which present the viewer with two offset, orthogonally polarised images of whatever is viewed through them. It was this effect that provided the first discovery of polarization, by Erasmus Bartholinus in 1669. In addition, the phase shift, and thus the change in polarization state, is usually frequency dependent, which, in combination with dichroism, often gives rise to bright colours and rainbow-like effects. In mineralogy, such properties, known as pleochroism, are frequently exploited for the purpose of identifying minerals using polarization microscopes. Additionally, many plastics that are not normally birefringent will become so when subject to mechanical stress, a phenomenon which is the basis of photoelasticity. Non-birefringent methods, to rotate the linear polarization of light beams, include the use of prismatic polarization rotators which use total internal reflection in a prism set designed for efficient collinear transmission.

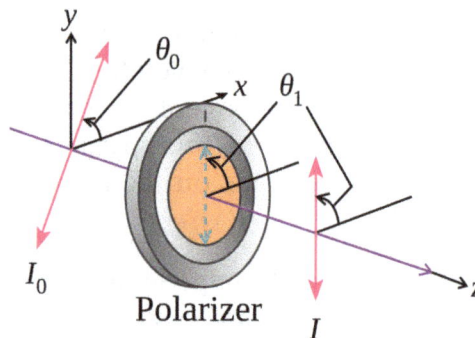

A polariser changing the orientation of linearly polarised light.
In this picture, $\theta_1 - \theta_0 = \theta_i$.

Media that reduce the amplitude of certain polarization modes are called *dichroic*, with devices that block nearly all of the radiation in one mode known as *polarizing filters* or simply "polarisers". Malus' law, which is named after Étienne-Louis Malus, says that when a perfect polariser is placed in a linear polarised beam of light, the intensity, I, of the light that passes through is given by

$$I = I_0 \cos^2 \theta_i \quad ,$$

where

$\quad$ I_0 is the initial intensity,

and θ_i is the angle between the light's initial polarization direction and the axis of the polariser.

A beam of unpolarised light can be thought of as containing a uniform mixture of linear polarizations at all possible angles. Since the average value of $\cos^2 \theta$ is 1/2, the transmission coefficient becomes

$$\frac{I}{I_0} = \frac{1}{2}$$

In practice, some light is lost in the polariser and the actual transmission of unpolarised light will be somewhat lower than this, around 38% for Polaroid-type polarisers but considerably higher (>49.9%) for some birefringent prism types.

In addition to birefringence and dichroism in extended media, polarization effects can also occur at the (reflective) interface between two materials of different refractive index. These effects are treated by the Fresnel equations. Part of the wave is transmitted and part is reflected, with the ratio depending on angle of incidence and the angle of refraction. In this way, physical optics recovers Brewster's angle. When light reflects from a thin film on a surface, interference between the reflections from the film's surfaces can produce polarization in the reflected and transmitted light.

Natural Light

The effects of a polarising filter on the sky in a photograph. Left picture is taken without polariser. For the right picture, filter was adjusted to eliminate certain polarizations of the scattered blue light from the sky.

Most sources of electromagnetic radiation contain a large number of atoms or molecules that emit light. The orientation of the electric fields produced by these emitters may not be correlated, in which case the light is said to be *unpolarised*. If there is partial correlation between the emitters, the light is *partially polarised*. If the polarization is consistent across the spectrum of the source, partially polarised light can be described as a superposition of a completely unpolarised component, and a completely polarised one. One may then describe the light in terms of the degree of polarization, and the parameters of the polarization ellipse.

Light reflected by shiny transparent materials is partly or fully polarised, except when the light is normal (perpendicular) to the surface. It was this effect that allowed the mathematician Étienne-Louis Malus to make the measurements that allowed for his development of the first mathematical models for polarised light. Polarization occurs when light is scattered in the atmosphere. The scattered light produces the brightness and colour in clear skies. This partial polarization of scattered light can be taken advantage of using polarizing filters to darken the sky in photographs. Optical polarization is principally of importance in chemistry due to circular dichroism and optical rotation (*"circular birefringence"*) exhibited by optically active (chiral) molecules.

Modern Optics

Modern optics encompasses the areas of optical science and engineering that became popular in the 20th century. These areas of optical science typically relate to the electromagnetic or quantum properties of light but do include other topics. A major subfield of modern optics, quantum optics, deals with specifically quantum mechanical properties of light. Quantum optics is not just theoretical; some modern devices, such as lasers, have principles of operation that depend on quantum mechanics. Light detectors, such as photomultipliers and channeltrons, respond to individual photons. Electronic image sensors, such as CCDs, exhibit shot noise corresponding to the statistics of individual photon events. Light-emitting diodes and photovoltaic cells, too, cannot be understood without quantum mechanics. In the study of these devices, quantum optics often overlaps with quantum electronics.

Specialty areas of optics research include the study of how light interacts with specific materials as in crystal optics and metamaterials. Other research focuses on the phenomenology of electromagnetic waves as in singular optics, non-imaging optics, non-linear optics, statistical optics, and radiometry. Additionally, computer engineers have taken an interest in integrated optics, machine vision, and photonic computing as possible components of the "next generation" of computers.

Today, the pure science of optics is called optical science or optical physics to distinguish it from applied optical sciences, which are referred to as optical engineering. Prominent subfields of optical engineering include illumination engineering, photonics, and optoelectronics with practical applications like lens design, fabrication and testing of optical components, and image processing. Some of these fields overlap, with nebulous boundaries between the subjects terms that mean slightly different things in different parts of the world and in different areas of industry. A professional community of researchers in nonlinear optics has developed in the last several decades due to advances in laser technology.

Lasers

A laser is a device that emits light (electromagnetic radiation) through a process called *stimulated emission*. The term *laser* is an acronym for *Light Amplification by Stimu-*

lated Emission of Radiation. Laser light is usually spatially coherent, which means that the light either is emitted in a narrow, low-divergence beam, or can be converted into one with the help of optical components such as lenses. Because the microwave equivalent of the laser, the *maser*, was developed first, devices that emit microwave and radio frequencies are usually called *masers.*

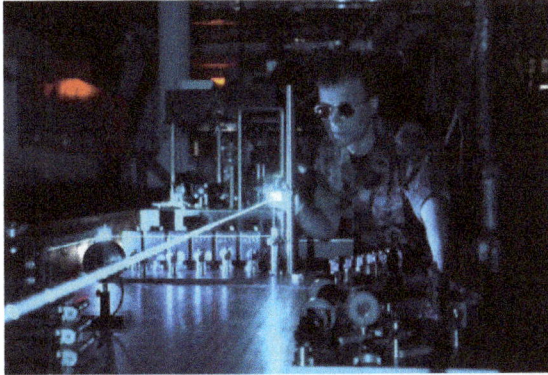

Experiments such as this one with high-power lasers are part of the modern optics research.

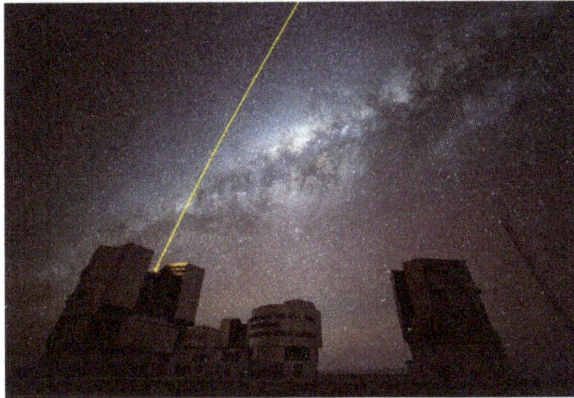

VLT's laser guided star.

The first working laser was demonstrated on 16 May 1960 by Theodore Maiman at Hughes Research Laboratories. When first invented, they were called "a solution looking for a problem". Since then, lasers have become a multibillion-dollar industry, finding utility in thousands of highly varied applications. The first application of lasers visible in the daily lives of the general population was the supermarket barcode scanner, introduced in 1974. The laserdisc player, introduced in 1978, was the first successful consumer product to include a laser, but the compact disc player was the first laser-equipped device to become truly common in consumers' homes, beginning in 1982. These optical storage devices use a semiconductor laser less than a millimetre wide to scan the surface of the disc for data retrieval. Fibre-optic communication relies on lasers to transmit large amounts of information at the speed of light. Other common applications of lasers include laser printers and laser pointers. Lasers are used in medicine in areas such as bloodless surgery, laser eye surgery, and laser capture microdissection and in military applications such as missile defence systems, electro-optical countermeasures (EOCM), and lidar. Lasers are also used in holograms, bubblegrams, laser light shows, and laser hair removal.

Kapitsa–Dirac Effect

The Kapitsa–Dirac effect causes beams of particles to diffract as the result of meeting a standing wave of light. Light can be used to position matter using various phenomena.

Applications

Optics is part of everyday life. The ubiquity of visual systems in biology indicates the central role optics plays as the science of one of the five senses. Many people benefit from eyeglasses or contact lenses, and optics are integral to the functioning of many consumer goods including cameras. Rainbows and mirages are examples of optical phenomena. Optical communication provides the backbone for both the Internet and modern telephony.

Human Eye

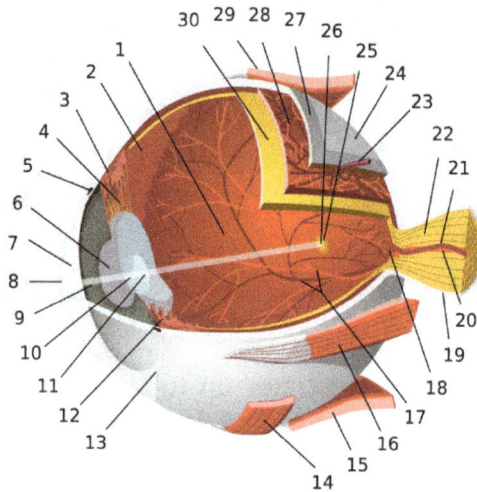

Model of a human eye. Features mentioned in this article are 3. ciliary muscle, 6. pupil, 8. cornea, 10. lens cortex, 22. optic nerve, 26. fovea, 30. retina

The human eye functions by focusing light onto a layer of photoreceptor cells called the retina, which forms the inner lining of the back of the eye. The focusing is accomplished by a series of transparent media. Light entering the eye passes first through the cornea, which provides much of the eye's optical power. The light then continues through the fluid just behind the cornea—the anterior chamber, then passes through the pupil. The light then passes through the lens, which focuses the light further and allows adjustment of focus. The light then passes through the main body of fluid in the eye—the vitreous humour, and reaches the retina. The cells in the retina line the back of the eye, except for where the optic nerve exits; this results in a blind spot.

oIn contrast, cone cells are less sensitive to the overall intensity of light, but come in three varieties that are sensitive to different frequency-ranges and thus are used in the perception of colour and photopic vision. Cone cells are highly concentrated in the fovea and have a high visual acuity meaning that they are better at spatial resolution than rod cells. Since cone cells are not as sensitive to dim light as rod cells, most night vision is limited to rod cells. Likewise, since cone cells are in the fovea, central vision (including the vision needed to do most reading, fine detail work such as sewing, or careful examination of objects) is done by cone cells.

Ciliary muscles around the lens allow the eye's focus to be adjusted. This process is known as accommodation. The near point and far point define the nearest and farthest distances from the eye at which an object can be brought into sharp focus. For a person with normal vision, the far point is located at infinity. The near point's location depends on how much the muscles can increase the curvature of the lens, and how inflexible the lens has become with age. Optometrists, ophthalmologists, and opticians usually consider an appropriate near point to be closer than normal reading distance—approximately 25 cm.

Defects in vision can be explained using optical principles. As people age, the lens becomes less flexible and the near point recedes from the eye, a condition known as presbyopia. Similarly, people suffering from hyperopia cannot decrease the focal length of their lens enough to allow for nearby objects to be imaged on their retina. Conversely, people who cannot increase the focal length of their lens enough to allow for distant objects to be imaged on the retina suffer from myopia and have a far point that is considerably closer than infinity. A condition known as astigmatism results when the cornea is not spherical but instead is more curved in one direction. This causes horizontally extended objects to be focused on different parts of the retina than vertically extended objects, and results in distorted images.

All of these conditions can be corrected using corrective lenses. For presbyopia and hyperopia, a converging lens provides the extra curvature necessary to bring the near point closer to the eye while for myopia a diverging lens provides the curvature necessary to send the far point to infinity. Astigmatism is corrected with a cylindrical surface lens that curves more strongly in one direction than in another, compensating for the non-uniformity of the cornea.

The optical power of corrective lenses is measured in diopters, a value equal to the reciprocal of the focal length measured in metres; with a positive focal length corresponding to a converging lens and a negative focal length corresponding to a diverging lens. For lenses that correct for astigmatism as well, three numbers are given: one for the spherical power, one for the cylindrical power, and one for the angle of orientation of the astigmatism.

Visual Effects

Optical illusions (also called visual illusions) are characterized by visually perceived images that differ from objective reality. The information gathered by the eye is processed in the brain to give a percept that differs from the object being imaged. Optical illusions can be the result of a variety of phenomena including physical effects that create images that are different from the objects that make them, the physiological effects on the eyes and brain of excessive stimulation (e.g. brightness, tilt, colour, movement), and cognitive illusions where the eye and brain make unconscious inferences.

The Ponzo Illusion relies on the fact that parallel lines appear to converge as they approach infinity.

Cognitive illusions include some which result from the unconscious misapplication of certain optical principles. For example, the Ames room, Hering, Müller-Lyer, Orbison, Ponzo, Sander, and Wundt illusions all rely on the suggestion of the appearance of distance by using converging and diverging lines, in the same way that parallel light rays (or indeed any set of parallel lines) appear to converge at a vanishing point at infinity in two-dimensionally rendered images with artistic perspective. This suggestion is also responsible for the famous moon illusion where the moon, despite having essentially the same angular size, appears much larger near the horizon than it does at zenith. This illusion so confounded Ptolemy that he incorrectly attributed it to atmospheric refraction when he described it in his treatise, *Optics*.

Another type of optical illusion exploits broken patterns to trick the mind into perceiving symmetries or asymmetries that are not present. Examples include the café wall, Ehrenstein, Fraser spiral, Poggendorff, and Zöllner illusions. Related, but not strictly illusions, are patterns that occur due to the superimposition of periodic structures. For example, transparent tissues with a grid structure produce shapes known as moiré patterns, while the superimposition of periodic transparent patterns comprising parallel opaque lines or curves produces line moiré patterns.

Optical Instruments

Single lenses have a variety of applications including photographic lenses, corrective lenses, and magnifying glasses while single mirrors are used in parabolic reflectors and rear-view mirrors. Combining a number of mirrors, prisms, and lenses produces compound optical instruments which have practical uses. For example, a periscope is simply two plane mirrors aligned to allow for viewing around obstructions. The most famous compound optical instruments in science are the microscope and the telescope which were both invented by the Dutch in the late 16th century.

Microscopes were first developed with just two lenses: an objective lens and an eyepiece. The objective lens is essentially a magnifying glass and was designed with a very small focal length while the eyepiece generally has a longer focal length. This has the effect of producing magnified images of close objects. Generally, an additional source of illumination is used since magnified images are dimmer due to the conservation of energy and the spreading of light rays over a larger surface area. Modern microscopes, known as *compound microscopes* have many lenses in them (typically four) to optimize the functionality and enhance image stability. A slightly different variety of microscope, the comparison microscope, looks at side-by-side images to produce a stereoscopic binocular view that appears three dimensional when used by humans.

The first telescopes, called *refracting telescopes* were also developed with a single objective and eyepiece lens. In contrast to the microscope, the objective lens of the telescope was designed with a large focal length to avoid optical aberrations. The objective focuses an image of a distant object at its focal point which is adjusted to be at the focal point of an eyepiece of a much smaller focal length. The main goal of a telescope is not necessarily magnification, but rather collection of light which is determined by the physical size of the objective lens. Thus, telescopes are normally indicated by the diameters of their objectives rather than by the magnification which can be changed by switching eyepieces. Because the magnification of a telescope is equal to the focal length of the objective divided by the focal length of the eyepiece, smaller focal-length eyepieces cause greater magnification.

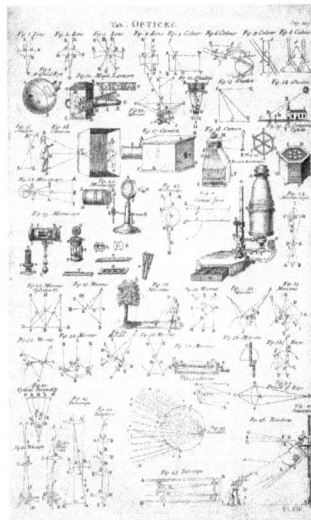

Illustrations of various optical instruments from the 1728 Cyclopaedia

Since crafting large lenses is much more difficult than crafting large mirrors, most modern telescopes are *reflecting telescopes*, that is, telescopes that use a primary mirror rather than an objective lens. The same general optical considerations apply to reflecting telescopes that applied to refracting telescopes, namely, the larger the primary mirror, the more light collected, and the magnification is still equal to the focal length of the primary mirror divided by the focal length of the eyepiece. Professional telescopes generally do not have eyepieces and instead place an instrument (often a charge-coupled device) at the focal point instead.

Photography

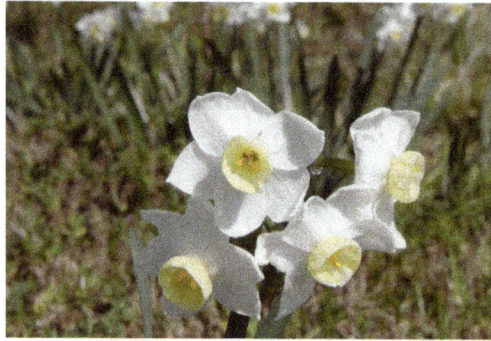

Photograph taken with aperture f/32

Photograph taken with aperture f/5

The optics of photography involves both lenses and the medium in which the electromagnetic radiation is recorded, whether it be a plate, film, or charge-coupled device. Photographers must consider the reciprocity of the camera and the shot which is summarized by the relation

$$\text{Exposure} \propto \text{ApertureArea} \times \text{ExposureTime} \times \text{SceneLuminance}$$

In other words, the smaller the aperture (giving greater depth of focus), the less light coming in, so the length of time has to be increased (leading to possible blurriness if motion occurs). An example of the use of the law of reciprocity is the Sunny 16 rule which gives a rough estimate for the settings needed to estimate the proper exposure in daylight.

A camera's aperture is measured by a unitless number called the f-number or f-stop, $f/\#$, often notated as N, and given by

$$f/\# = N = \frac{f}{D}$$

where f is the focal length, and D is the diameter of the entrance pupil. By convention, "$f/\#$" is treated as a single symbol, and specific values of $f/\#$ are written by replacing the number sign with the value. The two ways to increase the f-stop are to either decrease the diameter of the entrance pupil or change to a longer focal length (in the case of a zoom lens, this can be done by simply adjusting the lens). Higher f-numbers also have a larger depth of field due to the lens approaching the limit of a pinhole camera which is able to focus all images perfectly, regardless of distance, but requires very long exposure times.

The field of view that the lens will provide changes with the focal length of the lens. There are three basic classifications based on the relationship to the diagonal size of the film or sensor size of the camera to the focal length of the lens:

- Normal lens: angle of view of about 50° (called *normal* because this angle considered roughly equivalent to human vision) and a focal length approximately equal to the diagonal of the film or sensor.

- Wide-angle lens: angle of view wider than 60° and focal length shorter than a normal lens.

- Long focus lens: angle of view narrower than a normal lens. This is any lens with a focal length longer than the diagonal measure of the film or sensor. The most common type of long focus lens is the telephoto lens, a design that uses a special *telephoto group* to be physically shorter than its focal length.

Modern zoom lenses may have some or all of these attributes.

The absolute value for the exposure time required depends on how sensitive to light the medium being used is (measured by the film speed, or, for digital media, by the quantum efficiency). Early photography used media that had very low light sensitivity, and so exposure times had to be long even for very bright shots. As technology has improved, so has the sensitivity through film cameras and digital cameras.

Other results from physical and geometrical optics apply to camera optics. For example, the maximum resolution capability of a particular camera set-up is determined by the diffraction limit associated with the pupil size and given, roughly, by the Rayleigh criterion.

Atmospheric Optics

The unique optical properties of the atmosphere cause a wide range of spectacular optical phenomena. The blue colour of the sky is a direct result of Rayleigh scattering which redirects higher frequency (blue) sunlight back into the field of view of the observer. Because blue light is scattered more easily than red light, the sun takes on a reddish hue when it is observed through a thick atmosphere, as during a sunrise or sunset. Additional particulate matter in the sky can scatter different colours at different angles creating colourful glowing skies at dusk and dawn. Scattering off of ice crystals and other particles in the atmosphere are responsible for halos, afterglows, coronas, rays of sunlight, and sun dogs. The variation in these kinds of phenomena is due to different particle sizes and geometries.

A colourful sky is often due to scattering of light off particulates and pollution, as in this photograph of a sunset during the October 2007 California wildfires.

Mirages are optical phenomena in which light rays are bent due to thermal variations in the refraction index of air, producing displaced or heavily distorted images of distant objects. Other dramatic optical phenomena associated with this include the Novaya Zemlya effect where the sun appears to rise earlier than predicted with a distorted shape. A spectacular form of refraction occurs with a temperature inversion called the Fata Morgana where objects on the horizon or even beyond the horizon, such as islands, cliffs, ships or icebergs, appear elongated and elevated, like "fairy tale castles".

Rainbows are the result of a combination of internal reflection and dispersive refraction of light in raindrops. A single reflection off the backs of an array of raindrops produces a rainbow with an angular size on the sky that ranges from 40° to 42° with red on the outside. Double rainbows are produced by two internal reflections with angular size of 50.5° to 54° with violet on the outside. Because rainbows are seen with the sun 180° away from the centre of the rainbow, rainbows are more prominent the closer the sun is to the horizon.

History of Optics

Optics began with the development of lenses by the ancient Egyptians and Mesopotamians, followed by theories on light and vision developed by ancient Greek philosophers, and the development of geometrical optics in the Greco-Roman world. Optics was significantly reformed by the developments in the medieval Islamic world, such as the beginnings of physical and physiological optics, and then significantly advanced in early modern Europe, where diffractive optics began. These earlier studies on optics are now known as "classical optics". The term "modern optics" refers to areas of optical research that largely developed in the 20th century, such as wave optics and quantum optics.

Early History of Optics

The earliest known lenses were made from polished crystal, often quartz, and have been dated as early as 750 BC for Assyrian lenses such as the Nimrud / Layard lens. There are many similar lenses from ancient Egypt, Greece and Babylon. The ancient Romans and Greeks filled glass spheres with water to make lenses. However, glass lenses were not thought of until the Middle Ages.

Some lenses fixed in ancient Egyptian statues are much older than those mentioned above. There is some doubt as to whether or not they qualify as lenses, but they are undoubtedly glass and served at least ornamental purposes. The statues appear to be anatomically correct schematic eyes.

In ancient India, the philosophical schools of Samkhya and Vaisheshika, from around the 6th–5th century BC, developed theories on light. According to the Samkhya school, light is one of the five fundamental "subtle" elements (*tanmatra*) out of which emerge the gross elements.

In contrast, the Vaisheshika school gives an atomic theory of the physical world on the non-atomic ground of ether, space and time. The basic atoms are those of earth (*prthivı*), water (*apas*), fire (*tejas*), and air (*vayu*), that should not be confused with the ordinary meaning of these terms. These atoms are taken to form binary molecules that combine further to form larger molecules. Motion is defined in terms of the movement of the physical atoms. Light rays are taken to be a stream of high velocity of *tejas* (fire) atoms. The particles of light can exhibit different characteristics depending on the speed and the arrangements of the *tejas* atoms. Around the first century BC, the *Vishnu Purana* refers to sunlight as "the seven rays of the sun".

In the fifth century BC, Empedocles postulated that everything was composed of four elements; fire, air, earth and water. He believed that Aphrodite made the human eye out of the four elements and that she lit the fire in the eye which shone out from the eye making sight possible. If this were true, then one could see during the night just as well as during the day, so Empedocles postulated an interaction between rays from the eyes and rays from a source such as the sun.

In his *Optics* Greek mathematician Euclid observed that "things seen under a greater angle appear greater, and those under a lesser angle less, while those under equal angles appear equal". In the 36 propositions that follow, Euclid relates the apparent size of an object to its distance from the eye and investigates the apparent shapes of cylinders and cones when viewed from different angles. Pappus believed these results to be important in astronomy and included Euclid's *Optics*, along with his *Phaenomena*, in the *Little Astronomy*, a compendium of smaller works to be studied before the *Syntaxis* (*Almagest*) of Ptolemy.

In 55 BC, Lucretius, a Roman who carried on the ideas of earlier Greek atomists, wrote:

The light and heat of the sun; these are composed of minute atoms which, when they are shoved off, lose no time in shooting right across the interspace of air in the direction imparted by the shove.

— *Lucretius, On the nature of the Universe*

Despite being similar to later particle theories of light, Lucretius's views were not generally accepted and light was still theorized as emanating from the eye.

In his *Catoptrica*, Hero of Alexandria showed by a geometrical method that the actual path taken by a ray of light reflected from a plane mirror is shorter than any other reflected path that might be drawn between the source and point of observation.

In the second century Claudius Ptolemy, an Alexandrian Greek or Hellenized Egyptian, undertook studies of reflection and refraction. He measured the angles of refraction between air, water, and glass, and his published results indicate that he adjusted his measurements to fit his (incorrect) assumption that the angle of refraction is proportional to the angle of incidence.

The Indian Buddhists, such as Dignāga in the 5th century and Dharmakirti in the 7th century, developed a type of atomism that is a philosophy about reality being composed of atomic entities that are momentary flashes of light or energy. They viewed light as being an atomic entity equivalent to energy, similar to the modern concept of photons, though they also viewed all matter as being composed of these light/energy particles.

The Beginnings of Geometrical Optics

The early writers discussed here treated vision more as a geometrical than as a physical, physiological, or psychological problem. The first known author of a treatise on geometrical optics was the geometer Euclid (c. 325 BC–265 BC). Euclid began his study of optics as he began his study of geometry, with a set of self-evident axioms.

1. Lines (or visual rays) can be drawn in a straight line to the object.

2. Those lines falling upon an object form a cone.

3. Those things upon which the lines fall are seen.

4. Those things seen under a larger angle appear larger.

5. Those things seen by a higher ray, appear higher.

6. Right and left rays appear right and left.

7. Things seen within several angles appear clearer.

Euclid did not define the physical nature of these visual rays but, using the principles of geometry, he discussed the effects of perspective and the rounding of things seen at a distance.

Where Euclid had limited his analysis to simple direct vision, Hero of Alexandria (c. AD 10–70) extended the principles of geometrical optics to consider problems of reflection (catoptrics). Unlike Euclid, Hero occasionally commented on the physical nature of visual rays, indicating that they proceeded at great speed from the eye to the object seen and were reflected from smooth surfaces but could become trapped in the porosities of unpolished surfaces. This has come to be known as *emission theory.*

Hero demonstrated the equality of the angle of incidence and reflection on the grounds that this is the shortest path from the object to the observer. On this basis, he was able to define the fixed relation between an object and its image in a plane mirror. Specifically, the image appears to be as far behind the mirror as the object really is in front of the mirror.

Like Hero, Ptolemy (c. 90–c. 168) considered the visual rays as proceeding from the eye to the object seen, but, unlike Hero, considered that the visual rays were not discrete lines, but formed a continuous cone. Ptolemy extended the study of vision beyond direct and reflected vision; he also studied vision by refracted rays (dioptrics), when we see objects through the interface between two media of different density. He conducted experiments to measure the path of vision when we look from air to water, from air to glass, and from water to glass and tabulated the relationship between the incident and refracted rays.

His tabulated results have been studied for the air water interface, and in general the values he obtained reflect the theoretical refraction given by modern theory, but the outliers are distorted to represent Ptolemy's *a priori* model of the nature of refraction.

Optics and Vision in the Islamic World

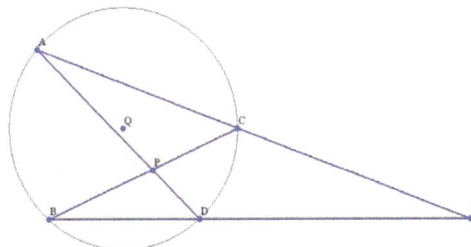

The theorem of Ibn Haytham

Al-Kindi (c. 801–873) was one of the earliest important optical writers in the Islamic world. In a work known in the west as *De radiis stellarum*, al-Kindi developed a theory "that everything in the world ... emits rays in every direction, which fill the whole world."

This theory of the active power of rays had an influence on later scholars such as Ibn al-Haytham, Robert Grosseteste and Roger Bacon.

Ibn Sahl (c. 940-1000) was an Iraqi mathematician associated with the court of Baghdad. About 984 he wrote a treatise *On Burning Mirrors and Lenses* in which he set out his understanding of how curved mirrors and lenses bend and focus light. In his work he discovered a law of refraction mathematically equivalent to Snell's law. He used his law of refraction to compute the shapes of lenses and mirrors that focus light at a single point on the axis.

Ibn al-Haytham (known in as *Alhacen* or *Alhazen* in Western Europe) (965–1040) produced a comprehensive and systematic analysis of Greek optical theories. Ibn al-Haytham's key achievement was twofold: first, to insist that vision occurred because of rays entering the eye; the second was to define the physical nature of the rays discussed by earlier geometrical optical writers, considering them as the forms of light and color. He then analyzed these physical rays according to the principles of geometrical optics. He wrote many books on optics, most significantly the *Book of Optics* (*Kitab al Manazir* in Arabic), translated into Latin as the *De aspectibus* or *Perspectiva*, which disseminated his ideas to Western Europe and had great influence on the later developments of optics.

Avicenna (980-1037) agreed with Alhazen that the speed of light is finite, as he "observed that if the perception of light is due to the emission of some sort of particles by a luminous source, the speed of light must be finite." Abū Rayhān al-Bīrūnī (973-1048) also agreed that light has a finite speed, and he was the first to discover that the speed of light is much faster than the speed of sound.

Abu ʻAbd Allah Muhammad ibn Maʼudh, who lived in Al-Andalus during the second half of the 11th century, wrote a work on optics later translated into Latin as *Liber de crepisculis*, which was mistakenly attributed to Alhazen. This was a "short work containing an estimation of the angle of depression of the sun at the beginning of the morning twilight and at the end of the evening twilight, and an attempt to calculate on the basis of this and other data the height of the atmospheric moisture responsible for the refraction of the sun's rays." Through his experiments, he obtained the value of 18°, which comes close to the modern value.

In the late 13th and early 14th centuries, Qutb al-Din al-Shirazi (1236–1311) and his student Kamāl al-Dīn al-Fārisī (1260–1320) continued the work of Ibn al-Haytham, and they were the first to give the correct explanations for the rainbow phenomenon. Al-Fārisī published his findings in his *Kitab Tanqih al-Manazir* (*The Revision of* [Ibn al-Haytham's] *Optics*).

Optics in Medieval Europe

The English bishop, Robert Grosseteste (c. 1175–1253), wrote on a wide range of scientific topics at the time of the origin of the medieval university and the recovery of the works of Aristotle. Grosseteste reflected a period of transition between the Platonism of early medieval learning and the new Aristotelianism, hence he tended to apply mathematics and the Platonic metaphor of light in many of his writings. He has been credited with discussing light from four different perspectives: an epistemology of light, a metaphysics or cosmogony of light, an etiology or physics of light, and a theology of light.

Setting aside the issues of epistemology and theology, Grosseteste's cosmogony of light describes the origin of the universe in what may loosely be described as a medieval "big bang" theory. Both his biblical commentary, the *Hexaemeron* (1230 x 35), and his scientific *On Light* (1235 x 40), took their inspiration from Genesis 1:3, "God said, let there be light", and described the subsequent process of creation as a natural physical process arising from the generative power of an expanding (and contracting) sphere of light.

His more general consideration of light as a primary agent of physical causation appears in his *On Lines, Angles, and Figures* where he asserts that "a natural agent propagates its power from itself to the recipient" and in *On the Nature of Places* where he notes that "every natural action is varied in strength and weakness through variation of lines, angles and figures."

Optical diagram showing light being refracted by a spherical glass container full of water. (from Roger Bacon, De multiplicatione specierum)

The English Franciscan, Roger Bacon (c. 1214–1294) was strongly influenced by Grosseteste's writings on the importance of light. In his optical writings (the *Perspectiva*, the *De multiplicatione specierum*, and the *De speculis comburentibus*) he cited a wide range of recently translated optical and philosophical works, including those of Alhacen, Aristotle, Avicenna, Averroes, Euclid, al-Kindi, Ptolemy, Tideus, and Constantine the African. Although he was not a slavish imitator, he drew his mathematical analysis of light and vision from the writings of the Arabic writer, Alhacen. But he added to this the Neoplatonic concept, perhaps drawn from Grosseteste, that every object radiates a power (*species*) by which it acts upon nearby objects suited to receive those *species*. Note that Bacon's optical use of the term "*species*" differs significantly from the genus / species categories found in Aristotelian philosophy.

Another English Franciscan, John Pecham (died 1292) built on the work of Bacon, Grosseteste, and a diverse range of earlier writers to produce what became the most widely used textbook on Optics of the Middle Ages, the *Perspectiva communis*. His book centered on the question of vision, on how we see, rather than on the nature of light and color. Pecham followed the model set forth by Alhacen, but interpreted Alhacen's ideas in the manner of Roger Bacon.

Like his predecessors, Witelo (c. 1230–1280 x 1314) drew on the extensive body of optical works recently translated from Greek and Arabic to produce a massive presentation of the subject entitled the *Perspectiva*. His theory of vision follows Alhacen and he does not consider Bacon's concept of *species*, although passages in his work demonstrate that he was influenced by Bacon's ideas. Judging from the number of surviving manuscripts, his work was not as influential as those of Pecham and Bacon, yet his importance, and that of Pecham, grew with the invention of printing.

- Peter of Limoges (1240–1306)

- Theodoric of Freiberg (ca. 1250–ca. 1310)

Renaissance and Early Modern Optics

Johannes Kepler (1571–1630) picked up the investigation of the laws of optics from his lunar essay of 1600. Both lunar and solar eclipses presented unexplained phenomena, such as unexpected shadow sizes, the red color of a total lunar eclipse, and the reportedly unusual light surrounding a total solar eclipse. Related issues of atmospheric refraction applied to all astronomical observations. Through most of 1603, Kepler paused his other work to focus on optical theory; the resulting manuscript, presented to the emperor on January 1, 1604, was published as *Astronomiae Pars Optica* (*The Optical Part of Astronomy*). In it, Kepler described the inverse-square law governing the intensity of light, reflection by flat and curved mirrors, and principles of pinhole cameras, as well as the astronomical implications of optics such as parallax and the apparent sizes of heavenly bodies. *Astronomiae Pars Optica* is generally recognized as the foundation of modern optics (though the law of refraction is conspicuously absent).

Willebrord Snellius (1580–1626) found the mathematical law of refraction, now known as Snell's law, in 1621. Subsequently René Descartes (1596–1650) showed, by using geometric construction and the law of refraction (also known as Descartes' law), that the angular radius of a rainbow is 42° (i.e. the angle subtended at the eye by the edge of the rainbow and the rainbow's centre is 42°). He also independently discovered the law of reflection, and his essay on optics was the first published mention of this law.

Christiaan Huygens (1629–1695) wrote several works in the area of optics. These included the *Opera reliqua* (also known as *Christiani Hugenii Zuilichemii, dum viveret Zelhemii toparchae, opuscula posthuma*) and the *Traité de la lumière*.

Isaac Newton (1643–1727) investigated the refraction of light, demonstrating that a prism could decompose white light into a spectrum of colours, and that a lens and a second prism could recompose the multicoloured spectrum into white light. He also showed that the coloured light does not change its properties by separating out a coloured beam and shining it on various objects. Newton noted that regardless of whether it was reflected or scattered or transmitted, it stayed the same colour. Thus, he observed that colour is the result of objects interacting with already-coloured light rather than objects generating the colour themselves. This is known as Newton's theory of colour. From this work he concluded that any refracting telescope would suffer from the dispersion of light into colours, and invented a reflecting telescope (today known as a Newtonian telescope) to bypass that problem. By grinding his own mirrors, using Newton's rings to judge the quality of the optics for his telescopes, he was able to produce a superior instrument to the refracting telescope, due primarily to the wider diameter of the mirror. In 1671 the Royal Society asked for a demonstration of his reflecting telescope. Their interest encouraged him to publish his notes *On Colour*, which he later expanded into his *Opticks*. Newton argued that light is composed of particles or *corpuscles* and were refracted by accelerating toward the denser medium, but he had to associate them with waves to explain the diffraction of light (*Opticks* Bk. II, Props. XII-L). Later physicists instead favoured a purely wavelike explanation of light to account for diffraction. Today's quantum mechanics, photons and the idea of wave-particle duality bear only a minor resemblance to Newton's understanding of light.

In his *Hypothesis of Light* of 1675, Newton posited the existence of the ether to transmit forces between particles. In 1704, Newton published *Opticks*, in which he expounded his corpuscular theory of light. He considered light to be made up of extremely subtle corpuscles, that ordinary matter was made of grosser corpuscles and speculated that through a kind of alchemical transmutation "Are not gross Bodies and Light convertible into one another, ...and may not Bodies receive much of their Activity from the Particles of Light which enter their Composition?"

The Beginnings of Diffractive Optics

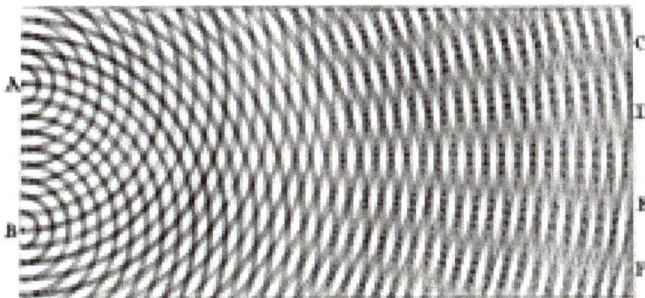

Thomas Young's sketch of two-slit diffraction, which he presented to the Royal Society in 1803

The effects of diffraction of light were first carefully observed and characterized by Francesco Maria Grimaldi, who also coined the term *diffraction*, from the Latin *diffrin-*

gere, 'to break into pieces', referring to light breaking up into different directions. The results of Grimaldi's observations were published posthumously in 1665. Isaac Newton studied these effects and attributed them to *inflexion* of light rays. James Gregory (1638–1675) observed the diffraction patterns caused by a bird feather, which was effectively the first diffraction grating. In 1803 Thomas Young did his famous experiment observing interference from two closely spaced slits in his double slit interferometer. Explaining his results by interference of the waves emanating from the two different slits, he deduced that light must propagate as waves. Augustin-Jean Fresnel did more definitive studies and calculations of diffraction, published in 1815 and 1818, and thereby gave great support to the wave theory of light that had been advanced by Christiaan Huygens and reinvigorated by Young, against Newton's particle theory.

Lenses and Lensmaking

The earliest known lenses were made from polished crystal, often quartz, and have been dated as early as 750 BC for Assyrian lenses such as the Layard / Nimrud lens. There are many similar lenses from ancient Egypt, Greece and Babylon. The ancient Romans and Greeks filled glass spheres with water to make lenses.

The earliest historical reference to magnification dates back to ancient Egyptian hieroglyphs in the 5th century BC, which depict "simple glass meniscal lenses". The earliest written record of magnification dates back to the 1st century AD, when Seneca the Younger, a tutor of Emperor Nero, wrote: "Letters, however small and indistinct, are seen enlarged and more clearly through a globe or glass filled with water". Emperor Nero is also said to have watched the gladiatorial games using an emerald as a corrective lens.

Ibn al-Haytham (Alhacen) wrote about the effects of pinhole, concave lenses, and magnifying glasses in his *Book of Optics*. Roger Bacon used parts of glass spheres as magnifying glasses and recommended them to be used to help people read. Roger Bacon got his inspiration from Alhacen in the 11th century. He discovered that light reflects from objects and does not get released from them. Around 1284 in Italy, Salvino D'Armate is credited with inventing the first wearable eye glasses.

Between the 11th and 13th century "reading stones" were invented. Often used by monks to assist in illuminating manuscripts, these were primitive plano-convex lenses initially made by cutting a glass sphere in half. As the stones were experimented with, it was slowly understood that shallower lenses magnified more effectively.

The earliest known working telescopes were the refracting telescopes that appeared in the Netherlands in 1608. Their development is credited to three individuals: Hans Lippershey and Zacharias Janssen, who were spectacle makers in Middelburg, and Jacob Metius of Alkmaar. Galileo greatly improved upon these designs the following year. Isaac Newton is credited with constructing the first functional reflecting telescope in 1668, his Newtonian reflector.

The first microscope was made around 1595 in Middelburg in the Dutch Republic. Three different eyeglass makers have been given credit for the invention: Hans Lippershey (who also developed the first real telescope); Hans Janssen; and his son, Zacharias. The coining of the name "microscope" has been credited to Giovanni Faber, who gave that name to Galileo Galilei's compound microscope in 1625.

Quantum Optics

Light is made up of particles called photons and hence inherently is quantized. Quantum optics is the study of the nature and effects of light as quantized photons. The first indication that light might be quantized came from Max Planck in 1899 when he correctly modelled blackbody radiation by assuming that the exchange of energy between light and matter only occurred in discrete amounts he called quanta. It was unknown whether the source of this discreteness was the matter or the light.In 1905, Albert Einstein published the theory of the photoelectric effect. It appeared that the only possible explanation for the effect was the quantization of light itself. Later, Niels Bohr showed that atoms could only emit discrete amounts of energy. The understanding of the interaction between light and matter following from these developments not only formed the basis of quantum optics but also were crucial for the development of quantum mechanics as a whole. However, the subfields of quantum mechanics dealing with matter-light interaction were principally regarded as research into matter rather than into light and hence, one rather spoke of atom physics and quantum electronics.

This changed with the invention of the maser in 1953 and the laser in 1960. Laser science—research into principles, design and application of these devices—became an important field, and the quantum mechanics underlying the laser's principles was studied now with more emphasis on the properties of light, and the name *quantum optics* became customary.

As laser science needed good theoretical foundations, and also because research into these soon proved very fruitful, interest in quantum optics rose. Following the work of Dirac in quantum field theory, George Sudarshan, Roy J. Glauber, and Leonard Mandel applied quantum theory to the electromagnetic field in the 1950s and 1960s to gain a more detailed understanding of photodetection and the statistics of light. This led to the introduction of the coherent state as a quantum description of laser light and the realization that some states of light could not be described with classical waves. In 1977, Kimble et al. demonstrated the first source of light which required a quantum description: a single atom that emitted one photon at a time. Another quantum state of light with certain advantages over any classical state, squeezed light, was soon proposed. At the same time, development of short and ultrashort laser pulses—created by Q-switching and mode-locking techniques—opened the way to the study of unimaginably fast ("ultrafast") processes. Applications for solid state research (e.g. Raman spectroscopy) were found, and mechanical forces of light on matter were studied. The latter led to levitating and positioning clouds of atoms or even small biological samples in an optical trap or optical tweezers by laser beam. This, along with Doppler cooling was the crucial technology needed to achieve the celebrated Bose–Einstein condensation.

Other remarkable results are the demonstration of quantum entanglement, quantum teleportation, and (recently, in 1995) quantum logic gates. The latter are of much interest in quantum information theory, a subject which partly emerged from quantum optics, partly from theoretical computer science.

Today's fields of interest among quantum optics researchers include parametric down-conversion, parametric oscillation, even shorter (attosecond) light pulses, use of quantum optics for quantum information, manipulation of single atoms, Bose–Einstein condensates, their application, and how to manipulate them (a sub-field often called atom optics), and much more.

Essential Concepts of Optics

The major concepts of optics are discussed in this chapter. It focuses on concepts such as reflection, refraction, diffraction, scattering etc. which are crucial for understanding optical devices. The following text elucidates the crucial concepts of optics, and strategically incorporates all the essential components of optics.

Reflection (Physics)

Reflection is the change in direction of a wavefront at an interface between two different media so that the wavefront returns into the medium from which it originated. Common examples include the reflection of light, sound and water waves. The *law of reflection* says that for specular reflection the angle at which the wave is incident on the surface equals the angle at which it is reflected. Mirrors exhibit specular reflection.

The reflection of Mount Hood in Mirror Lake.

In acoustics, reflection causes echoes and is used in sonar. In geology, it is important in the study of seismic waves. Reflection is observed with surface waves in bodies of water. Reflection is observed with many types of electromagnetic wave, besides visible light. Reflection of VHF and higher frequencies is important for radio transmission and for radar. Even hard X-rays and gamma rays can be reflected at shallow angles with special "grazing" mirrors.

Reflection of Light

Reflection of light is either *specular* (mirror-like) or *diffuse* (retaining the energy, but losing the image) depending on the nature of the interface. In specular reflection the phase of the reflected waves depends on the choice of the origin of coordinates, but the relative phase between s and p (TE and TM) polarizations is fixed by the properties of the media and of the interface between them.

A mirror provides the most common model for specular light reflection, and typically consists of a glass sheet with a metallic coating where the reflection actually occurs. Reflection is enhanced in metals by suppression of wave propagation beyond their skin depths. Reflection also occurs at the surface of transparent media, such as water or glass.

In the diagram at left, a light ray PO strikes a vertical mirror at point O, and the reflected ray is OQ. By projecting an imaginary line through point O perpendicular to the mirror, known as the *normal*, we can measure the *angle of incidence*, θ_i and the *angle of reflection*, θ_r. The *law of reflection* states that $\theta_i = \theta_r$, *or in other words, the angle of incidence equals the angle of reflection.*

In fact, reflection of light may occur whenever light travels from a medium of a given refractive index into a medium with a different refractive index. In the most general case, a certain fraction of the light is reflected from the interface, and the remainder is refracted. Solving Maxwell's equations for a light ray striking a boundary allows the derivation of the Fresnel equations, which can be used to predict how much of the light is reflected, and how much is refracted in a given situation. This is analogous to the way impedance mismatch in an electric circuit causes reflection of signals. Total internal reflection of light from a denser medium occurs if the angle of incidence is above the critical angle.

Total internal reflection is used as a means of focusing waves that cannot effectively be reflected by common means. X-ray telescopes are constructed by creating a converging "tunnel" for the waves. As the waves interact at low angle with the surface of this tunnel they are reflected toward the focus point (or toward another interaction with the tunnel surface, eventually being directed to the detector at the focus). A conventional reflector would be useless as the X-rays would simply pass through the intended reflector.

When light reflects off a material denser (with higher refractive index) than the external medium, it undergoes a polarity inversion. In contrast, a less dense, lower refractive index material will reflect light in phase. This is an important principle in the field of thin-film optics.

Specular reflection forms images. Reflection from a flat surface forms a mirror image, which appears to be reversed from left to right because we compare the image we see to what we would see if we were rotated into the position of the image. Specular reflection at a curved surface forms an image which may be magnified or demagnified; curved mirrors have optical power. Such mirrors may have surfaces that are spherical or parabolic.

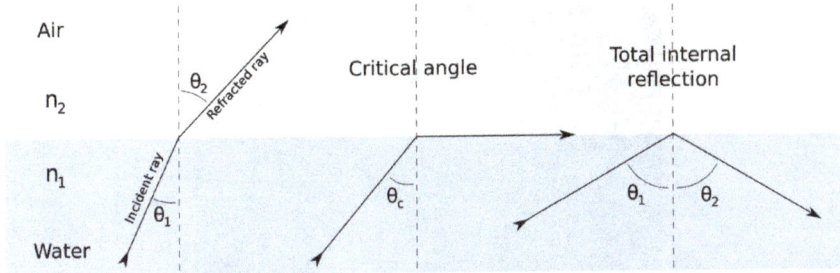

Refraction of light at the interface between two media.

Laws of Reflection

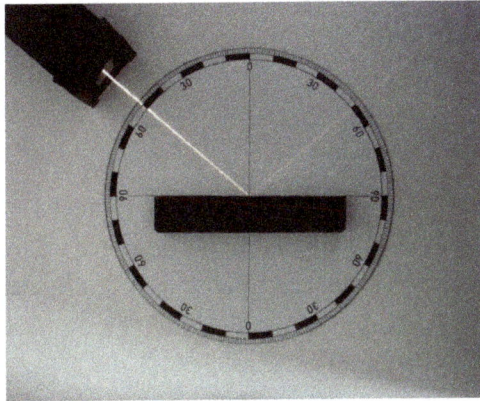

An example of the law of reflection

If the reflecting surface is very smooth, the reflection of light that occurs is called specular or regular reflection. The laws of reflection are as follows:

1. The incident ray, the reflected ray and the normal to the reflection surface at the point of the incidence lie in the same plane.

2. The angle which the incident ray makes with the normal is equal to the angle which the reflected ray makes to the same normal.

3. The reflected ray and the incident ray are on the opposite sides of the normal.

These three laws can all be derived from the Fresnel equations.

Mechanism

In classical electrodynamics, light is considered as an electromagnetic wave, which is described by Maxwell's equations. Light waves incident on a material induce small oscillations of polarisation in the individual atoms (or oscillation of electrons, in metals), causing each particle to radiate a small secondary wave in all directions, like a dipole antenna. All these waves add up to give specular reflection and refraction, according to the Huygens–Fresnel principle.

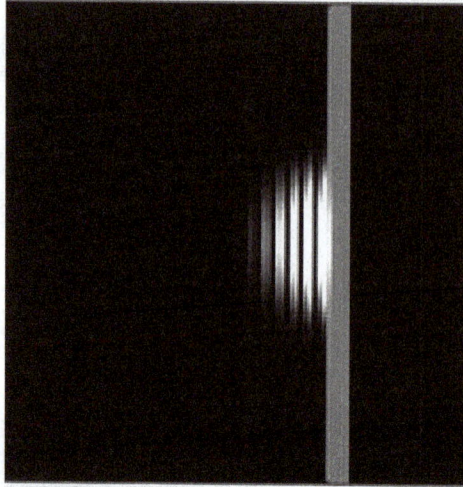

2D simulation: reflection of a quantum particle. White blur represents the probability distribution of finding a particle in a given place if measured.

In the case of dielectrics such as glass, the electric field of the light acts on the electrons in the material, and the moving electrons generate fields and become new radiators. The refracted light in the glass is the combination of the forward radiation of the electrons and the incident light. The reflected light is the combination of the backward radiation of all of the electrons.

In metals, electrons with no binding energy are called free electrons. When these electrons oscillate with the incident light, the phase difference between their radiation field and the incident field is π (180°), so the forward radiation cancels the incident light, and backward radiation is just the reflected light.

Light–matter interaction in terms of photons is a topic of quantum electrodynamics, and is described in detail by Richard Feynman in his popular book *QED: The Strange Theory of Light and Matter*.

Diffuse Reflection

When light strikes the surface of a (non-metallic) material it bounces off in all directions due to multiple reflections by the microscopic irregularities *inside* the material (e.g. the grain boundaries of a polycrystalline material, or the cell or fiber boundaries of an organic material) and by its surface, if it is rough. Thus, an 'image' is not formed. This is called *diffuse reflection*. The exact form of the reflection depends on the structure of the material. One common model for diffuse reflection is Lambertian reflectance, in which the light is reflected with equal luminance (in photometry) or radiance (in radiometry) in all directions, as defined by Lambert's cosine law.

The light sent to our eyes by most of the objects we see is due to diffuse reflection from their surface, so that this is our primary mechanism of physical observation.

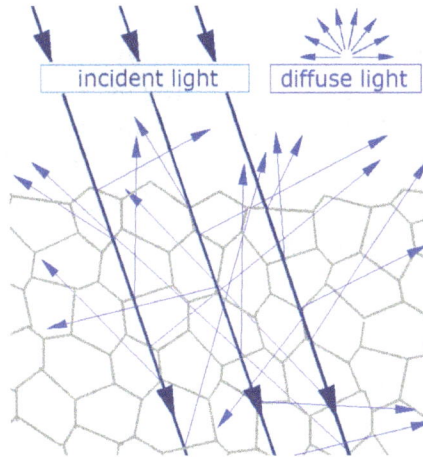

General scattering mechanism which gives diffuse reflection by a solid surface

Retroreflection

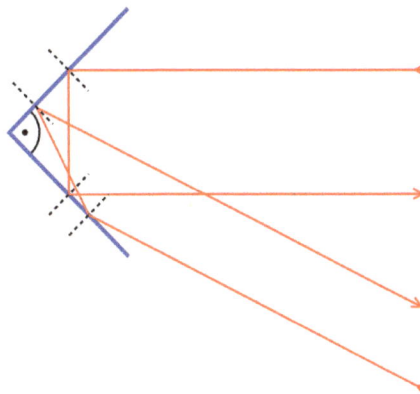

Working principle of a corner reflector

Some surfaces exhibit *retroreflection*. The structure of these surfaces is such that light is returned in the direction from which it came.

When flying over clouds illuminated by sunlight the region seen around the aircraft's shadow will appear brighter, and a similar effect may be seen from dew on grass. This partial retro-reflection is created by the refractive properties of the curved droplet's surface and reflective properties at the backside of the droplet.

Some animals' retinas act as retroreflectors, as this effectively improves the animals' night vision. Since the lenses of their eyes modify reciprocally the paths of the incoming and outgoing light the effect is that the eyes act as a strong retroreflector, sometimes seen at night when walking in wildlands with a flashlight.

A simple retroreflector can be made by placing three ordinary mirrors mutually perpendicular to one another (a corner reflector). The image produced is the inverse of one produced by a single mirror. A surface can be made partially retroreflective by depositing a layer of tiny refractive spheres on it or by creating small pyramid like structures. In both cases internal reflection causes the light to be reflected back to where it originated. This is used to make traffic signs and automobile license plates reflect light mostly back in the direction from which it came. In this application perfect retroreflection is not desired, since the light would then be directed back into the headlights of an oncoming car rather than to the driver's eyes.

Multiple Reflections

When light reflects off a mirror, one image appears. Two mirrors placed exactly face to face give the appearance of an infinite number of images along a straight line. The multiple images seen between two mirrors that sit at an angle to each other lie over a circle. The center of that circle is located at the imaginary intersection of the mirrors. A square of four mirrors placed face to face give the appearance of an infinite number of images arranged in a plane. The multiple images seen between four mirrors assembling a pyramid, in which each pair of mirrors sits an angle to each other, lie over a sphere. If the base of the pyramid is rectangle shaped, the images spread over a section of a torus.

Complex Conjugate Reflection

In this process (which is also known as phase conjugation), light bounces exactly back in the direction from which it came due to a nonlinear optical process. Not only the direction of the light is reversed, but the actual wavefronts are reversed as well. A conjugate reflector can be used to remove aberrations from a beam by reflecting it and then passing the reflection through the aberrating optics a second time.

Other Types of Reflection

Neutron Reflection

Materials that reflect neutrons, for example beryllium, are used in nuclear reactors and nuclear weapons. In the physical and biological sciences, the reflection of neutrons off of atoms within a material is commonly used to determine the material's internal structure.

Sound Reflection

When a longitudinal sound wave strikes a flat surface, sound is reflected in a coherent manner provided that the dimension of the reflective surface is large compared to the wavelength of the sound. Note that audible sound has a very wide frequency range (from 20 to about 17000 Hz), and thus a very wide range of wavelengths (from about 20 mm to 17 m). As a result, the overall nature of the reflection varies according to the texture and structure of the surface. For example, porous materials will absorb some energy, and rough materials (where rough is relative to the wavelength) tend to reflect in many directions—to scatter the energy, rather than to reflect it coherently. This leads into the field of architectural acoustics, because the nature of these reflections is critical to the auditory feel of a space. In the theory of exterior noise mitigation, reflective surface size mildly detracts from the concept of a noise barrier by reflecting some of the sound into the opposite direction.

Sound diffusion panel for high frequencies

Seismic Reflection

Seismic waves produced by earthquakes or other sources (such as explosions) may be reflected by layers within the Earth. Study of the deep reflections of waves generated by earthquakes has allowed seismologists to determine the layered structure of the Earth. Shallower reflections are used in reflection seismology to study the Earth's crust generally, and in particular to prospect for petroleum and natural gas deposits.

Refraction

Refraction is the change in direction of propagation of a wave due to a change in its transmission medium.

The phenomenon is explained by the conservation of energy and the conservation of momentum. Due to the change of medium, the phase velocity of the wave is changed but its frequency remains constant. This is most commonly observed when a wave passes from one medium to another at any angle other than 0° from the normal. Refraction of light is the most commonly observed phenomenon, but any type of wave can refract when it interacts with a medium, for example when sound waves pass from one medium into another or when water waves move into water of a different depth. Refraction is described by Snell's law, which states that, for a given pair of media and a wave with a single frequency, the ratio of the sines of the angle of incidence θ_1 and angle of refraction θ_2 is equivalent to the ratio of phase velocities (v_1 / v_2) in the two media, or equivalently, to the opposite ratio of the indices of refraction (n_2 / n_1):

$$\frac{\sin \theta_1}{\sin \theta_2} = \frac{v_1}{v_2} = \frac{n_2}{n_1}.$$

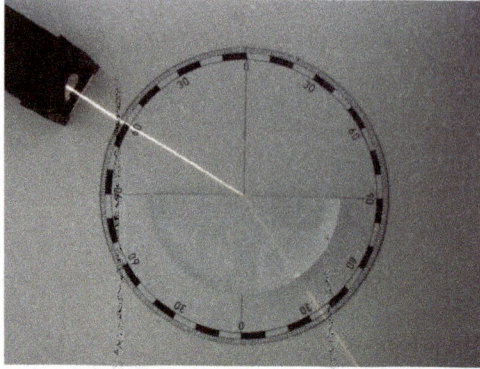

Light on air–plexi surface in this experiment undergoes refraction (lower ray) and reflection (upper ray).

Refraction in a glass of water. The image is flipped.

An image of the Golden Gate Bridge is refracted and bent by many differing three-dimensional drops of water.

In general, the incident wave is partially refracted and partially reflected; the details of this behavior are described by the Fresnel equations.

Explanation

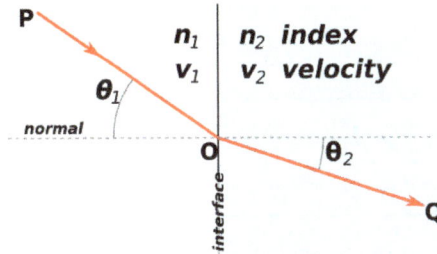

Refraction of light at the interface between two media of different refractive indices, with n2 > n1. Since the phase velocity is lower in the second medium (v2 < v1), the angle of refraction θ2 is less than the angle of incidence θ1; that is, the ray in the higher-index medium is closer to the normal.

In optics, refraction is a phenomenon that often occurs when waves travel from a medium with a given refractive index to a medium with another at an oblique angle. At the boundary between the media, the wave's phase velocity is altered, usually causing a change in direction. Its wavelength increases or decreases, but its frequency remains constant. For example, a light ray will refract as it enters and leaves glass, assuming there is a change in refractive index. A ray traveling along the normal (perpendicular to the boundary) will change speed, but not direction. Refraction still occurs in this case. Understanding of this concept led to the invention of lenses and the refracting telescope.

Refraction can be seen when looking into a bowl of water. Air has a refractive index of about 1.0003, and water has a refractive index of about 1.3330. If a person looks at a straight object, such as a pencil or straw, which is placed at a slant, partially in the water, the object appears to bend at the water's surface. This is due to the bending of light rays as they move from the water to the air. Once the rays reach the eye, the eye traces them back as straight lines (lines of sight). The lines of sight (shown as dashed lines) intersect at a higher position than where the actual rays originated. This causes the pencil to appear higher and the water to appear shallower than it really is. The depth that the water appears to be when viewed from above is known as the *apparent depth*. This is an important consideration for spearfishing from the surface because it will make the target fish appear to be in a different place, and the fisher must aim lower to catch the fish. Conversely, an object above the water has a higher *apparent height* when viewed from below the water. The opposite correction must be made by an archer fish. For small angles of incidence (measured from the normal, when $\sin \theta$ is approximately the same as $\tan \theta$), the ratio of apparent to real depth is the ratio of the refractive indexes of air to that of water. But, as the angle of incidence approaches 90°, the apparent depth approaches zero, albeit reflection increases, which limits observation at high angles of incidence. Conversely, the apparent height approaches infinity as the angle of incidence (from below) increases, but even earlier, as the angle of total internal reflection is approached, albeit the image also fades from view as this limit is approached.

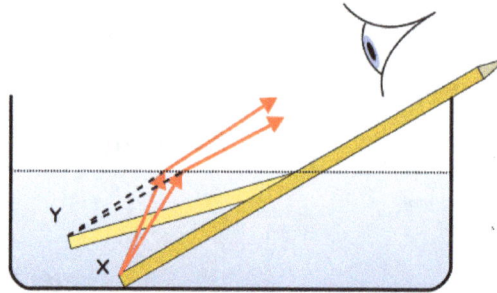

An object (in this case a pencil) part immersed in water looks bent due to refraction: the light waves from X change direction and so seem to originate at Y. (More accurately, for any angle of view, Y should be vertically above X, and the pencil should appear shorter, not longer as shown.)

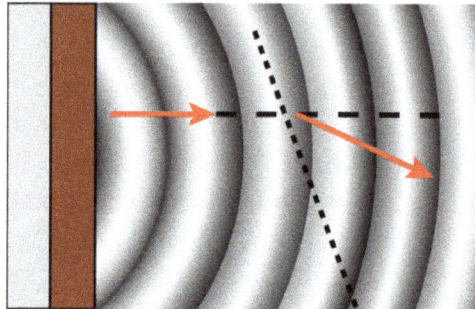

Diagram of refraction of water waves

The diagram on the right shows an example of refraction in water waves. Ripples travel from the left and pass over a shallower region inclined at an angle to the wavefront. The waves travel slower in the more shallow water, so the wavelength decreases and the wave bends at the boundary. The dotted line represents the normal to the boundary. The dashed line represents the original direction of the waves. This phenomenon explains why waves on a shoreline tend to strike the shore close to a perpendicular angle. As the waves travel from deep water into shallower water near the shore, they are refracted from their original direction of travel to an angle more normal to the shoreline. Refraction is also responsible for rainbows and for the splitting of white light into a rainbow-spectrum as it passes through a glass prism. Glass has a higher refractive index than air. When a beam of white light passes from air into a material having an index of refraction that varies with frequency, a phenomenon known as dispersion occurs, in which different coloured components of the white light are refracted at different angles, i.e., they bend by different amounts at the interface, so that they become separated. The different colors correspond to different frequencies.

While refraction allows for phenomena such as rainbows, it may also produce peculiar optical phenomena, such as mirages and Fata Morgana. These are caused by the change of the refractive index of air with temperature.

The refractive index of materials can also be nonlinear, as occurs with the Kerr effect when high intensity light leads to a refractive index proportional to the intensity of the incident light.

Recently, some metamaterials have been created that have a negative refractive index. With metamaterials, we can also obtain total refraction phenomena when the wave impedances of the two media are matched. There is then no reflected wave.

Also, since refraction can make objects appear closer than they are, it is responsible for allowing water to magnify objects. First, as light is entering a drop of water, it slows down. If the water's surface is not flat, then the light will be bent into a new path. This round shape will bend the light outwards and as it spreads out, the image you see gets larger.

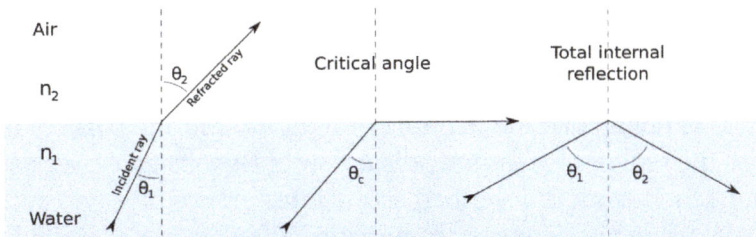

Refraction of light at the interface between two media.

An analogy that is often put forward to explain the refraction of light is as follows: "Imagine a marching band as it marches at an oblique angle from a pavement (a fast medium) into mud (a slower medium). The marchers on the side that runs into the mud first will slow down first. This causes the whole band to pivot slightly toward the normal (make a smaller angle from the normal)."

Why refraction occurs when light travels from a medium with a given refractive index to a medium with another, can be explained by the path integral formulation of quantum mechanics (the complete method was developed in 1948 by Richard Feynman). Feynman humorously explained it himself in the recording "QED: Fits of Reflection and Transmission - Quantum Behaviour - Richard Feynman (The Sir Douglas Robb Lectures, University of Auckland, 1979)".

Clinical Significance

In medicine, particularly optometry, ophthalmology and orthoptics, *refraction* (also known as *refractometry*) is a clinical test in which a phoropter may be used by the appropriate eye care professional to determine the eye's refractive error and the best corrective lenses to be prescribed. A series of test lenses in graded optical powers or focal lengths are presented to determine which provides the sharpest, clearest vision.

Acoustics

In underwater acoustics, refraction is the bending or curving of a sound ray that results when the ray passes through a sound speed gradient from a region of one sound speed to a region of a different speed. The amount of ray bending is dependent on the amount of difference between sound speeds, that is, the variation in temperature, salinity, and pressure of the water. Similar acoustics effects are also found in the Earth's atmosphere. The phenomenon of refraction of sound in the atmosphere has been known for centuries; however, beginning in the early 1970s, widespread analysis of this effect came into vogue through the designing of urban highways and noise barriers to address the meteorological effects of bending of sound rays in the lower atmosphere.

Diffraction

Diffraction refers to various phenomena which occur when a wave encounters an obstacle or a slit. It is defined as the bending of light around the corners of an obstacle or aperture into the region of geometrical shadow of the obstacle. In classical physics, the diffraction phenomenon is described as the interference of waves according to the Huygens–Fresnel principle. These characteristic behaviors are exhibited when a wave encounters an obstacle or a slit that is comparable in size to its wavelength. Similar effects occur when a light wave travels through a medium with a varying refractive index, or when a sound wave travels through a medium with varying acoustic impedance. Diffraction occurs with all waves, including sound waves, water waves, and electromagnetic waves such as visible light, X-rays and radio waves.

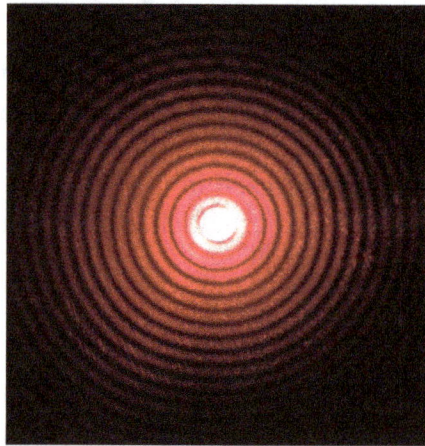

Diffraction pattern of red laser beam made on a plate after passing a small circular hole in another plate

Since physical objects have wave-like properties (at the atomic level), diffraction also occurs with matter and can be studied according to the principles of quantum mechanics. Italian scientist Francesco Maria Grimaldi coined the word "diffraction" and was the first to record accurate observations of the phenomenon in 1660.

While diffraction occurs whenever propagating waves encounter such changes, its effects are generally most pronounced for waves whose wavelength is roughly comparable to the dimensions of the diffracting object or slit. If the obstructing object provides multiple, closely spaced openings, a complex pattern of varying intensity can result. This is due to the addition, or interference, of different parts of a wave that travels to the observer by different paths, where different path lengths result in different phases. The formalism of diffraction can also describe the way in which waves of finite extent propagate in free space. For example, the expanding profile of a laser beam, the beam shape of a radar antenna and the field of view of an ultrasonic transducer can all be analyzed using diffraction equations.

Examples

The effects of diffraction are often seen in everyday life. The most striking examples of diffraction are those that involve light; for example, the closely spaced tracks on a CD or DVD act as a diffraction grating to form the familiar rainbow pattern seen when looking at a disk. This principle can be extended to engineer a grating with a structure such that it will produce any diffraction pattern desired; the hologram on a credit card is an example. Diffraction in the atmosphere by small particles can cause a bright ring to be visible around a bright light source like the sun or the moon. A shadow of a solid object, using light from a compact source, shows small fringes near its edges. The speckle pattern which is observed when laser light falls on an optically rough surface is also a diffraction phenomenon. When deli meat appears to be iridescent, that is diffraction off the meat fibers. All these effects are a consequence of the fact that light propagates as a wave.

Solar glory at the steam from hot springs. A glory is an optical phenomenon produced by light backscattered (a combination of diffraction, reflection and refraction) towards its source by a cloud of uniformly sized water droplets.

Diffraction can occur with any kind of wave. Ocean waves diffract around jetties and other obstacles. Sound waves can diffract around objects, which is why one can still hear someone calling even when hiding behind a tree. Diffraction can also be a concern in some technical applications; it sets a fundamental limit to the resolution of a camera, telescope, or microscope.

History

The law of diffraction of light, which would much later come to be known as Snell's law was first accurately described by the scientist Ibn Sahl at the Baghdad court in 984. In the manuscript *On Burning Mirrors and Lenses*, Sahl used the law to derive lens shapes that focus light with no geometric aberrations.

The effects of diffraction of light were first carefully observed and characterized by Francesco Maria Grimaldi, who also coined the term *diffraction*, from the Latin *diffringere*, 'to break into pieces', referring to light breaking up into different directions. The results of Grimaldi's observations were published posthumously in 1665. Isaac Newton studied these effects and attributed them to *inflexion* of light rays. James Gregory (1638–1675) observed the diffraction patterns caused by a bird feather, which was effectively the first diffraction grating to be discovered. Thomas Young performed a celebrated experiment in 1803 demonstrating interference from two closely spaced slits. Explaining his results by interference of the waves emanating from the two different slits, he deduced that light must propagate as waves. Augustin-Jean Fresnel did more definitive studies and calculations of diffraction, made public in 1815 and 1818, and thereby gave great support to the wave theory of light that had been advanced by Christiaan Huygens and reinvigorated by Young, against Newton's particle theory.

Mechanism

Photograph of single-slit diffraction in a circular ripple tank

Diffraction arises because of the way in which waves propagate; this is described by the Huygens–Fresnel principle and the principle of superposition of waves. The propagation of a wave can be visualized by considering every particle of the transmitted medium on a wavefront as a point source for a secondary spherical wave. The wave displacement at any subsequent point is the sum of these secondary waves. When waves are added together, their sum is determined by the relative phases as well as the amplitudes of the individual waves so that the summed amplitude of the waves can have any value between zero and the sum of the individual amplitudes. Hence, diffraction patterns usually have a series of maxima and minima.

There are various analytical models which allow the diffracted field to be calculated, including the Kirchhoff-Fresnel diffraction equation which is derived from wave equation, the Fraunhofer diffraction approximation of the Kirchhoff equation which applies to the far field and the Fresnel diffraction approximation which applies to the near field. Most configurations cannot be solved analytically, but can yield numerical solutions through finite element and boundary element methods.

It is possible to obtain a qualitative understanding of many diffraction phenomena by considering how the relative phases of the individual secondary wave sources vary, and in particular, the conditions in which the phase difference equals half a cycle in which case waves will cancel one another out.

The simplest descriptions of diffraction are those in which the situation can be reduced to a two-dimensional problem. For water waves, this is already the case; water waves propagate only on the surface of the water. For light, we can often neglect one direction if the diffracting object extends in that direction over a distance far greater than the wavelength. In the case of light shining through small circular holes we will have to take into account the full three-dimensional nature of the problem.

Diffraction of Light

Some examples of diffraction of light are considered below.

Single-slit Diffraction

Numerical approximation of diffraction pattern from a slit of width equal to wavelength of an incident plane wave in 3D spectrum visualization

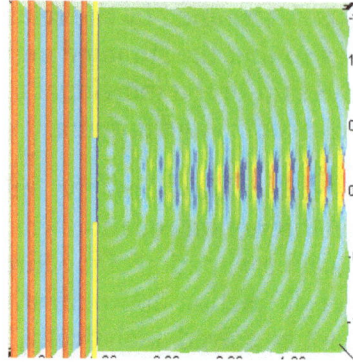

Numerical approximation of diffraction pattern from a slit of width equal to five times the wavelength of an incident plane wave in 3D spectrum visualization

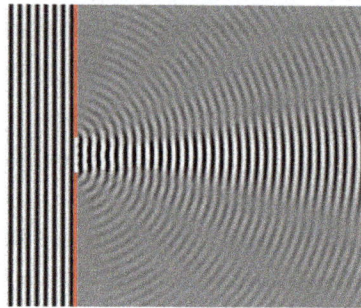

Numerical approximation of diffraction pattern from a slit of width four wavelengths with an incident plane wave. The main central beam, nulls, and phase reversals are apparent.

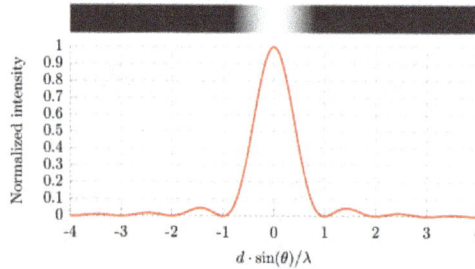

Graph and image of single-slit diffraction.

A long slit of infinitesimal width which is illuminated by light diffracts the light into a series of circular waves and the wavefront which emerges from the slit is a cylindrical wave of uniform intensity.

A slit which is wider than a wavelength produces interference effects in the space downstream of the slit. These can be explained by assuming that the slit behaves as though it has a large number of point sources spaced evenly across the width of the slit. The analysis of this system is simplified if we consider light of a single wavelength. If the incident light is coherent, these sources all have the same phase. Light incident at a given point in the space downstream of the slit is made up of contributions from each of these point sources and if the relative phases of these contributions vary by 2π or more, we may expect to find minima and maxima in the diffracted light. Such phase differences are caused by differences in the path lengths over which contributing rays reach the point from the slit.

We can find the angle at which a first minimum is obtained in the diffracted light by the following reasoning. The light from a source located at the top edge of the slit interferes destructively with a source located at the middle of the slit, when the path difference between them is equal to $\lambda/2$. Similarly, the source just below the top of the slit will interfere destructively with the source located just below the middle of the slit at the same angle. We can continue this reasoning along the entire height of the slit to conclude that the condition for destructive interference for the entire slit is the same as the condition for destructive interference between two narrow slits a distance apart that is half the width of the slit. The path difference is given by $\dfrac{d\sin(\theta)}{2}$ so that the minimum intensity occurs at an angle θ_{min} given by

$$d\sin\theta_{min} = \lambda$$

where

- d is the width of the slit,

- θ_{min} is the angle of incidence at which the minimum intensity occurs, and

- λ is the wavelength of the light

A similar argument can be used to show that if we imagine the slit to be divided into four, six, eight parts, etc., minima are obtained at angles θ_n given by

$$d\sin\theta_n = n\lambda$$

where

- n is an integer other than zero.

There is no such simple argument to enable us to find the maxima of the diffraction pattern. The intensity profile can be calculated using the Fraunhofer diffraction equation as

$$I(\theta) = I_0\,\text{sinc}^2\left(\frac{d\pi}{\lambda}\sin\theta\right)$$

where

- $I(\theta)$ is the intensity at a given angle,

- I_0 is the original intensity, and

- the unnormalized sinc function above is given by $\text{sinc}(x) = \dfrac{\sin x}{x}$ if $x \neq 0$, and $\text{sinc}(0)=1$

This analysis applies only to the far field, that is, at a distance much larger than the width of the slit.

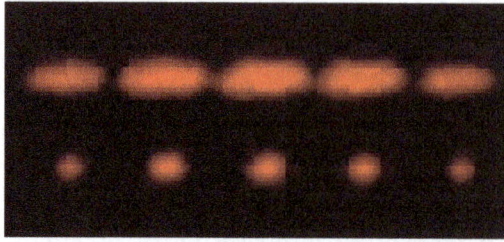

2-slit (top) and 5-slit diffraction of red laser light

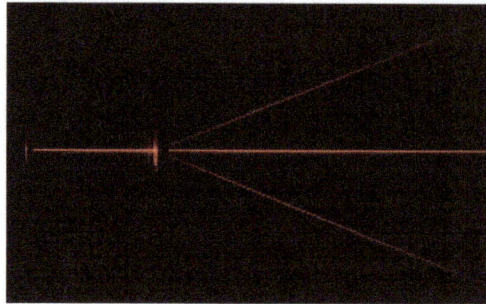

Diffraction of a red laser using a diffraction grating.

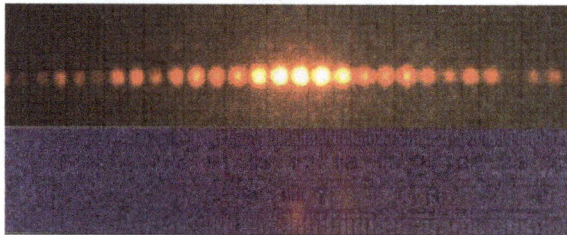

A diffraction pattern of a 633 nm laser through a grid of 150 slits

Diffraction Grating

A diffraction grating is an optical component with a regular pattern. The form of the light diffracted by a grating depends on the structure of the elements and the number of elements present, but all gratings have intensity maxima at angles θ_m which are given by the grating equation

$$d\left(\sin\theta_m + \sin\theta_i\right) = m\lambda.$$

where

- θ_i is the angle at which the light is incident,

- d is the separation of grating elements, and

- m is an integer which can be positive or negative.

The light diffracted by a grating is found by summing the light diffracted from each of the elements, and is essentially a convolution of diffraction and interference patterns.

The figure shows the light diffracted by 2-element and 5-element gratings where the grating spacings are the same; it can be seen that the maxima are in the same position, but the detailed structures of the intensities are different.

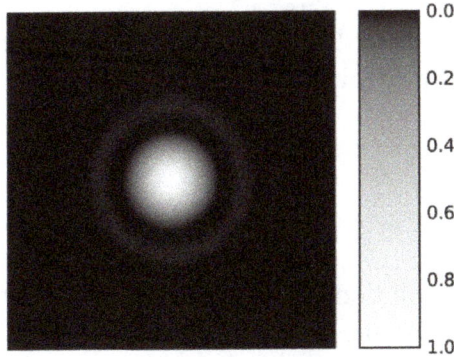

A computer-generated image of an Airy disk.

Computer generated light diffraction pattern from a circular aperture of diameter 0.5 micrometre at a wavelength of 0.6 micrometre (red-light) at distances of 0.1 cm – 1 cm in steps of 0.1 cm. One can seethe image moving from the Fresnel region into the Fraunhofer region where the Airy pattern is seen.

Circular Aperture

The far-field diffraction of a plane wave incident on a circular aperture is often referred to as the Airy Disk. The variation in intensity with angle is given by

$$I(\theta) = I_0 \left(\frac{2J_1(ka\sin\theta)}{ka\sin\theta} \right)^2,$$

where a is the radius of the circular aperture, k is equal to $2\pi/\lambda$ and J_1 is a Bessel function. The smaller the aperture, the larger the spot size at a given distance, and the greater the divergence of the diffracted beams.

General Aperture

The wave that emerges from a point source has amplitude ψ at location r that is given by the solution of the frequency domain wave equation for a point source (The Helmholtz Equation),

$$\nabla^2 \psi + k^2 \psi = \delta(\mathbf{r})$$

where $\delta(\mathbf{r})$ is the 3-dimensional delta function. The delta function has only radial dependence, so the Laplace operator (a.k.a. scalar Laplacian) in the spherical coordinate system simplifies to

$$\nabla^2 \psi = \frac{1}{r}\frac{\partial^2}{\partial r^2}(r\psi)$$

By direct substitution, the solution to this equation can be readily shown to be the scalar Green's function, which in the spherical coordinate system (and using the physics time convention $e^{-i\omega t}$) is:

$$\psi(r) = \frac{e^{ikr}}{4\pi r}$$

This solution assumes that the delta function source is located at the origin. If the source is located at an arbitrary source point, denoted by the vector $\mathbf{r}'$ and the field point is located at the point $\mathbf{r}$, then we may represent the scalar Green's function (for arbitrary source location) as:

$$\psi(\mathbf{r}\,|\,\mathbf{r}') = \frac{e^{ik|\mathbf{r}-\mathbf{r}'|}}{4\pi\,|\mathbf{r}-\mathbf{r}'|}$$

Therefore, if an electric field, $E_{inc}(x,y)$ is incident on the aperture, the field produced by this aperture distribution is given by the surface integral:

$$\Psi(r) \propto \iint\limits_{\text{aperture}} E_{inc}(x',y')\frac{e^{ik|\mathbf{r}-\mathbf{r}'|}}{4\pi\,|\mathbf{r}-\mathbf{r}'|}dx'\,dy',$$

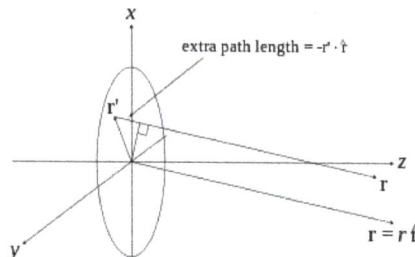

On the calculation of Fraunhofer region fields

where the source point in the aperture is given by the vector

$$\mathbf{r}' = x'\hat{\mathbf{x}} + y'\hat{\mathbf{y}}$$

In the far field, wherein the parallel rays approximation can be employed, the Green's function,

$$\psi(\mathbf{r}\,|\,\mathbf{r}') = \frac{e^{ik|\mathbf{r}-\mathbf{r}'|}}{4\pi\,|\mathbf{r}-\mathbf{r}'|}$$

simplifies to

$$\psi(\mathbf{r}\,|\,\mathbf{r}') = \frac{e^{ikr}}{4\pi r}e^{-ik(\mathbf{r}'\cdot\hat{\mathbf{r}})}$$

as can be seen in the figure to the right (click to enlarge).

The expression for the far-zone (Fraunhofer region) field becomes

$$\Psi(r) \propto \frac{e^{ikr}}{4\pi r} \iint\limits_{\text{aperture}} E_{\text{inc}}(x',y')e^{-ik(\mathbf{r}'\cdot\hat{\mathbf{r}})}\,dx'dy',$$

Now, since

$$\mathbf{r}' = x'\hat{\mathbf{x}} + y'\hat{\mathbf{y}}$$

and

$$\hat{\mathbf{r}} = \sin\theta\cos\phi\hat{\mathbf{x}} + \sin\theta\,\sin\phi\,\hat{\mathbf{y}} + \cos\theta\hat{\mathbf{z}}$$

the expression for the Fraunhofer region field from a planar aperture now becomes,

$$\Psi(r) \propto \frac{e^{ikr}}{4\pi r} \iint\limits_{\text{aperture}} E_{\text{inc}}(x',y')e^{-ik\sin\theta(\cos\phi x'+\sin\phi y')}\,dx'dy'$$

Letting,

$$k_x = k \sin \theta \cos \phi$$

and

$$k_y = k \sin \theta \sin \phi$$

the Fraunhofer region field of the planar aperture assumes the form of a Fourier transform

$$\Psi(r) \propto \frac{e^{ikr}}{4\pi r} \iint_{\text{aperture}} E_{\text{inc}}(x', y') e^{-i(k_x x' + k_y y')} \, dx' dy',$$

In the far-field / Fraunhofer region, this becomes the spatial Fourier transform of the aperture distribution. Huygens' principle when applied to an aperture simply says that the far-field diffraction pattern is the spatial Fourier transform of the aperture shape, and this is a direct by-product of using the parallel-rays approximation, which is identical to doing a plane wave decomposition of the aperture plane fields.

Propagation of a Laser Beam

The way in which the beam profile of a laser beam changes as it propagates is determined by diffraction. When the entire emitted beam has a planar, spatially coherent wave front, it approximates Gaussian beam profile and has the lowest divergence for a given diameter. The smaller the output beam, the quicker it diverges. It is possible to reduce the divergence of a laser beam by first expanding it with one convex lens, and then collimating it with a second convex lens whose focal point is coincident with that of the first lens. The resulting beam has a larger diameter, and hence a lower divergence. Divergence of a laser beam may be reduced below the diffraction of a Gaussian beam or even reversed to convergence if the refractive index of the propagation media increases with the light intensity. This may result in a self-focusing effect.

When the wave front of the emitted beam has perturbations, only the transverse coherence length (where the wave front perturbation is less than 1/4 of the wavelength) should be considered as a Gaussian beam diameter when determining the divergence of the laser beam. If the transverse coherence length in the vertical direction is higher than in horizontal, the laser beam divergence will be lower in the vertical direction than in the horizontal.

Diffraction-limited Imaging

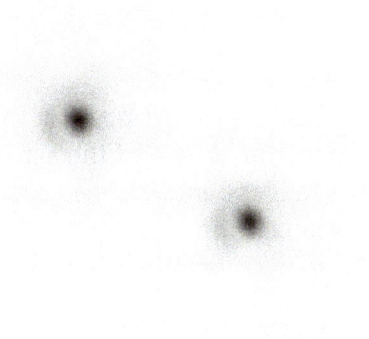

The Airy disk around each of the stars from the 2.56 m telescope aperture can be seen in this *lucky image* of the binary star zeta Boötis.

The ability of an imaging system to resolve detail is ultimately limited by diffraction. This is because a plane wave incident on a circular lens or mirror is diffracted as described above. The light is not focused to a point but forms an Airy disk having a central spot in the focal plane with radius to first null of

$$d = 1.22\lambda N,$$

where λ is the wavelength of the light and N is the f-number (focal length divided by diameter) of the imaging optics. In object space, the corresponding angular resolution is

$$\sin\theta = 1.22\frac{\lambda}{D},$$

where D is the diameter of the entrance pupil of the imaging lens (e.g., of a telescope's main mirror).

Two point sources will each produce an Airy pattern. As the point sources move closer together, the patterns will start to overlap, and ultimately they will merge to form a single pattern, in which case the two point sources cannot be resolved in the image. The Rayleigh criterion specifies that two point sources can be considered to be resolvable if the separation of the two images is at least the radius of the Airy disk, i.e. if the first minimum of one coincides with the maximum of the other.

Thus, the larger the aperture of the lens, and the smaller the wavelength, the finer the resolution of an imaging system. This is why telescopes have very large lenses or mirrors, and why optical microscopes are limited in the detail which they can see.

Speckle Patterns

The speckle pattern which is seen when using a laser pointer is another diffraction phenomenon. It is a result of the superposition of many waves with different phases, which are produced when a laser beam illuminates a rough surface. They add together to give a resultant wave whose amplitude, and therefore intensity, varies randomly.

Babinet's Principle

Babinet's Principle is a useful theorem stating that the diffraction pattern from an opaque body is identical to that from a hole of the same size and shape, but with differing intensities. This means that the interference conditions of a single obstruction would be the same as that of a single slit.

Patterns

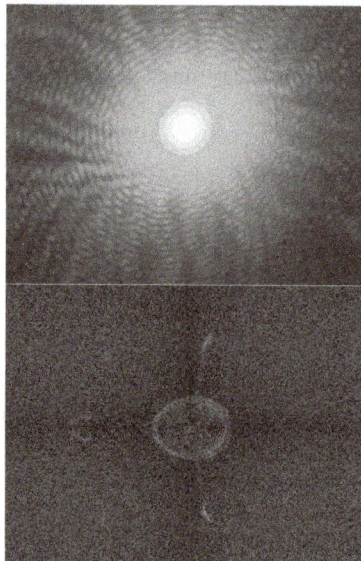

The upper half of this image shows a diffraction pattern of He-Ne laser beam on an elliptic aperture. The lower half is its 2D Fourier transform approximately reconstructing the shape of the aperture.

Several qualitative observations can be made of diffraction in general:

- The angular spacing of the features in the diffraction pattern is inversely proportional to the dimensions of the object causing the diffraction. In other words: The smaller the diffracting object, the 'wider' the resulting diffraction pattern, and vice versa. (More precisely, this is true of the sines of the angles.)

- The diffraction angles are invariant under scaling; that is, they depend only on the ratio of the wavelength to the size of the diffracting object.

- When the diffracting object has a periodic structure, for example in a diffraction grating, the features generally become sharper. The third figure, for example, shows a comparison of a double-slit pattern with a pattern formed by five slits, both sets of slits having the same spacing, between the center of one slit and the next.

Particle Diffraction

Quantum theory tells us that every particle exhibits wave properties. In particular, massive particles can interfere and therefore diffract. Diffraction of electrons and neutrons stood as one of the powerful arguments in favor of quantum mechanics. The wavelength associated with a particle is the de Broglie wavelength

$$\lambda = \frac{h}{p}$$

where h is Planck's constant and p is the momentum of the particle (mass × velocity for slow-moving particles).

For most macroscopic objects, this wavelength is so short that it is not meaningful to assign a wavelength to them. A sodium atom traveling at about 30,000 m/s would have a De Broglie wavelength of about 50 pico meters.

Because the wavelength for even the smallest of macroscopic objects is extremely small, diffraction of matter waves is only visible for small particles, like electrons, neutrons, atoms and small molecules. The short wavelength of these matter waves makes them ideally suited to study the atomic crystal structure of solids and large molecules like proteins.

Relatively larger molecules like buckyballs were also shown to diffract.

Bragg Diffraction

Diffraction from a three-dimensional periodic structure such as atoms in a crystal is called Bragg diffraction. It is similar to what occurs when waves are scattered from a diffraction grating. Bragg diffraction is a consequence of interference between waves reflecting from different crystal planes. The condition of constructive interference is given by *Bragg's law*:

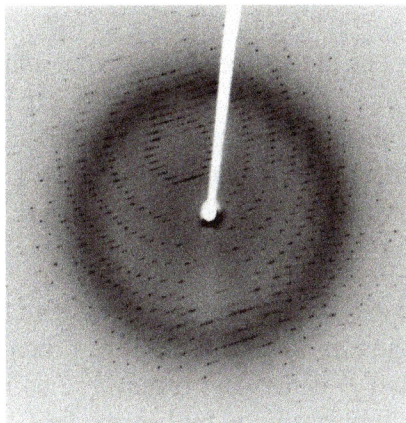

Following Bragg's law, each dot (or reflection) in this diffraction pattern forms from the constructive interference of X-rays passing through a crystal. The data can be used to determine the crystal's atomic structure.

$$m\lambda = 2d \sin \theta$$

where

λ is the wavelength,

d is the distance between crystal planes,

θ is the angle of the diffracted wave.

and m is an integer known as the *order* of the diffracted beam.

Bragg diffraction may be carried out using either light of very short wavelength like X-rays or matter waves like neutrons (and electrons) whose wavelength is on the order of (or much smaller than) the atomic spacing. The pattern produced gives information of the separations of crystallographic planes d, allowing one to deduce the crystal structure. Diffraction contrast, in electron microscopes and x-topography devices in particular, is also a powerful tool for examining individual defects and local strain fields in crystals.

Coherence

The description of diffraction relies on the interference of waves emanating from the same source taking different paths to the same point on a screen. In this description, the difference in phase between waves that took different paths is only dependent on the effective path length. This does not take into account the fact that waves that arrive at the screen at the same time were emitted by the source at different times. The initial phase with which the source emits waves can change over time in an unpredictable way. This means that waves emitted by the source at times that are too far apart can no longer form a constant interference pattern since the relation between their phases is no longer time independent.

The length over which the phase in a beam of light is correlated, is called the coherence length. In order for interference to occur, the path length difference must be smaller than the coherence length. This is sometimes referred to as spectral coherence, as it is related to the presence of different frequency components in the wave. In the case of light emitted by an atomic transition, the coherence length is related to the lifetime of the excited state from which the atom made its transition.

If waves are emitted from an extended source, this can lead to incoherence in the transversal direction. When looking at a cross section of a beam of light, the length over which the phase is correlated is called the transverse coherence length. In the case of Young's double slit experiment, this would mean that if the transverse coherence length is smaller than the spacing between the two slits, the resulting pattern on a screen would look like two single slit diffraction patterns.

In the case of particles like electrons, neutrons and atoms, the coherence length is related to the spatial extent of the wave function that describes the particle.

Scattering

Scattering is a general physical process where some forms of radiation, such as light, sound, or moving particles, are forced to deviate from a straight trajectory by one or more paths due to localized non-uniformities in the medium through which they pass. In conventional use, this also includes deviation of reflected radiation from the angle predicted by the law of reflection. Reflections that undergo scattering are often called *diffuse reflections* and unscattered reflections are called *specular* (mirror-like) reflections.

Scattering may also refer to particle-particle collisions between molecules, atoms, electrons, photons and other particles. Examples include: cosmic ray scattering in the Earth's upper atmosphere; particle collisions inside particle accelerators; electron scattering by gas atoms in fluorescent lamps; and neutron scattering inside nuclear reactors.

The types of non-uniformities which can cause scattering, sometimes known as *scatterers* or *scattering centers*, are too numerous to list, but a small sample includes particles, bubbles, droplets, density fluctuations in fluids, crystallites in polycrystalline solids, defects in monocrystalline solids, surface roughness, cells in organisms, and textile fibers in clothing. The effects of such features on the path of almost any type of propagating wave or moving particle can be described in the framework of scattering theory.

Some areas where scattering and scattering theory are significant include radar sensing, medical ultrasound, semiconductor wafer inspection, polymerization process monitoring, acoustic tiling, free-space communications and computer-generated imagery. Particle-particle scattering theory is important in areas such as particle physics, atomic, molecular, and optical physics, nuclear physics and astrophysics.

Single and Multiple Scattering

Zodiacal light is a glow originated in the scattering of sunlight by dust located between the planets that are spread through the plane of the Solar System.

When radiation is only scattered by one localized scattering center, this is called *single scattering*. It is very common that scattering centers are grouped together; in such cases, radiation may scatter many times, in what is known as *multiple scattering*. The main difference between the effects of single and multiple scattering is that single scattering can usually be treated as a random phenomenon, whereas multiple scattering is usually more stochastic. Because the location of a single scattering center is not usually well known relative to the path of the radiation, the outcome, which tends to depend strongly on the exact incoming trajectory, appears random to an observer. This type of scattering would be exemplified by an electron being fired at an atomic nucleus. In this case, the atom's exact position relative to the path of the electron is unknown and would be unmeasurable, so the exact trajectory of the electron after the collision cannot be predicted. Single scattering is therefore often described by probability distributions.

With multiple scattering, the randomness of the interaction tends to be averaged out by the large number of scattering events, so that the final path of the radiation appears to be a deterministic distribution of intensity. This is exemplified by a light beam passing through thick fog. Multiple scattering is highly analogous to diffusion, and the terms *multiple scattering* and *diffusion* are interchangeable in many contexts. Optical elements designed to produce multiple scattering are thus known as *diffusers*. Coherent backscattering, an enhancement of backscattering that occurs when coherent radiation is multiply scattered by a random medium, is usually attributed to weak localization.

Not all single scattering is random, however. A well-controlled laser beam can be exactly positioned to scatter off a microscopic particle with a deterministic outcome, for instance. Such situations are encountered in radar scattering as well, where the targets tend to be macroscopic objects such as people or aircraft.

Similarly, multiple scattering can sometimes have somewhat random outcomes, particularly with coherent radiation. The random fluctuations in the multiply scattered intensity of coherent radiation are called speckles. Speckle also occurs if multiple parts of a coherent wave scatter from different centers. In certain rare circumstances, multiple scattering may only involve a small number of interactions such that the randomness is not completely averaged out. These systems are considered to be some of the most difficult to model accurately.

The description of scattering and the distinction between single and multiple scattering are tightly related to wave–particle duality.

Scattering Theory

Scattering theory is a framework for studying and understanding the scattering of waves and particles. Prosaically, wave scattering corresponds to the collision and scattering of a wave with some material object, for instance sunlight scattered by rain drops to form a rainbow. Scattering also includes the interaction of billiard balls on a table, the Rutherford scattering (or angle change) of alpha particles by gold nuclei, the Bragg scattering (or diffraction) of electrons and X-rays by a cluster of atoms, and the inelastic scattering of a fission fragment as it traverses a thin foil. More precisely, scattering consists of the study of how solutions of partial differential equations, propagating freely "in the distant past", come together and interact with one another or with a boundary condition, and then propagate away "to the distant future".

Electromagnetic Scattering

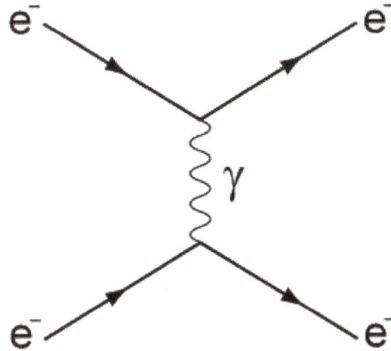

A Feynman diagram of scattering between two electrons by emission of a virtual photon.

Electromagnetic waves are one of the best known and most commonly encountered forms of radiation that undergo scattering. Scattering of light and radio waves (especially in radar) is particularly important. Several different aspects of electromagnetic scattering are distinct enough to have conventional names. Major forms of elastic light scattering (involving negligible energy transfer) are Rayleigh scattering and Mie scattering. Inelastic scattering includes Brillouin scattering, Raman scattering, inelastic X-ray scattering and Compton scattering.

Light scattering is one of the two major physical processes that contribute to the visible appearance of most objects, the other being absorption. Surfaces described as *white* owe their appearance to multiple scattering of light by internal or surface inhomogeneities in the object, for example by the boundaries of transparent microscopic crystals that make up a stone or by the microscopic fibers in a sheet of paper. More generally, the gloss (or lustre or sheen) of the surface is determined by scattering. Highly scattering surfaces are described as being dull or having a matte finish, while the absence of surface scattering leads to a glossy appearance, as with polished metal or stone.

Spectral absorption, the selective absorption of certain colors, determines the color of most objects with some modification by elastic scattering. The apparent blue color of veins in skin is a common example where both spectral absorption and scattering play important and complex roles in the coloration. Light scattering can also create color without absorption, often shades of blue, as with the sky (Rayleigh scattering), the human blue iris, and the feathers of some birds (Prum et al. 1998). However, resonant light scattering in nanoparticles can produce many different highly saturated and vibrant hues, especially when surface plasmon resonance is involved (Roqué et al. 2006).

Models of light scattering can be divided into three domains based on a dimensionless size parameter, α which is defined as:

$$\alpha = \pi D_p / \lambda,$$

where πD_p is the circumference of a particle and λ is the wavelength of incident radiation. Based on the value of α, these domains are:

$\alpha \ll 1$: Rayleigh scattering (small particle compared to wavelength of light);

$\alpha \approx 1$: Mie scattering (particle about the same size as wavelength of light, valid only for spheres);

$\alpha \gg 1$: geometric scattering (particle much larger than wavelength of light).

Rayleigh scattering is a process in which electromagnetic radiation (including light) is scattered by a small spherical volume of variant refractive index, such as a particle, bubble, droplet, or even a density fluctuation. This effect was first modeled successfully by Lord Rayleigh, from whom it gets its name. In order for Rayleigh's model to apply, the sphere must be much smaller in diameter than the wavelength (λ) of the scattered wave; typically the upper limit is taken to be about 1/10 the wavelength. In this size regime, the exact shape of the scattering center is usually not very significant and can often be treated as a sphere of equivalent volume. The inherent scattering that radiation undergoes passing through a pure gas is due to microscopic density fluctuations as the gas molecules move around, which are normally small enough in scale for Rayleigh's model to apply. This scattering mechanism is the primary cause of the blue color of the Earth's sky on a clear day, as the shorter blue wavelengths of sunlight passing overhead are more strongly scattered than the longer red wavelengths according to Rayleigh's famous $1/\lambda^4$ relation. Along with absorption, such scattering is a major cause of the attenuation of radiation by the atmosphere. The degree of scattering varies as a function of the ratio of the particle diameter to the wavelength of the radiation, along with many other factors including polarization, angle, and coherence.

For larger diameters, the problem of electromagnetic scattering by spheres was first solved by Gustav Mie, and scattering by spheres larger than the Rayleigh range is therefore usually known as Mie scattering. In the Mie regime, the shape of the scattering center becomes much more significant and the theory only applies well to spheres and, with some modification, spheroids and ellipsoids. Closed-form solutions for scattering by certain other simple shapes exist, but no general closed-form solution is known for arbitrary shapes.

Both Mie and Rayleigh scattering are considered elastic scattering processes, in which the energy (and thus wavelength and frequency) of the light is not substantially changed. However, electromagnetic radiation scattered by moving scattering centers does undergo a Doppler shift, which can be detected and used to measure the velocity of the scattering center/s in forms of techniques such as lidar and radar. This shift involves a slight change in energy.

At values of the ratio of particle diameter to wavelength more than about 10, the laws of geometric optics are mostly sufficient to describe the interaction of light with the particle, and at this point the interaction is not usually described as scattering.

For modeling of scattering in cases where the Rayleigh and Mie models do not apply such as irregularly shaped particles, there are many numerical methods that can be used. The most common are finite-element methods which solve Maxwell's equations to find the distribution of the scattered electromagnetic field. Sophisticated software packages exist which allow the user to specify the refractive index or indices of the scattering feature in space, creating a 2- or sometimes 3-dimensional model of the structure. For relatively large and complex structures, these models usually require substantial execution times on a computer.

Optical Resolution

Optical resolution describes the ability of an imaging system to resolve detail in the object that is being imaged.

An imaging system may have many individual components including a lens and recording and display components. Each of these contributes to the optical resolution of the system, as will the environment in which the imaging is done.

Lateral Resolution

Resolution depends on the distance between two distinguishable radiating points. The sections below describe the theoretical estimates of resolution, but the real values may differ. The results below are based on mathematical models of Airy discs, which assumes an adequate level of contrast. In low-contrast systems, the resolution may be much lower than predicted by the theory outlined below. Real optical systems are complex and practical difficulties often increase the distance between distinguishable point sources.

The resolution of a system is based on the minimum distance at which the points can be distinguished as individuals. Several standards are used to determine, quantitatively, whether or not the points can be distinguished. One of the methods specifies that, on the line between the center of one point and the next, the contrast between the maximum and minimum intensity be at least 26% lower than the maximum. This corresponds to the overlap of one airy disk on the first dark ring in the other. This standard for separation is also known as the Rayleigh criterion. In symbols, the distance is defined as follows:

$$r = \frac{1.22\lambda}{2n\sin\theta} = \frac{0.61\lambda}{\text{NA}}$$

where

r is the minimum distance between resolvable points, in the same units as is specified

λ is the wavelength of light, emission wavelength, in the case of fluorescence,

n is the index of refraction of the media surrounding the radiating points,

θ is the half angle of the pencil of light that enters the objective, and

NA is the numerical aperture

This formula is suitable for confocal microscopy, but is also used in traditional microscopy. In confocal laser-scanned microscopes, the full-width half half-maximum (FWHM) of the point spread function is often used to avoid the difficulty of measuring the Airy disc. This, combined with the rastered illumination pattern, results in better resolution, but it is still proportional to the Rayleigh-based formula given above.

$$r = \frac{0.4\lambda}{\text{NA}}$$

Also common in the microscopy literature is a formula for resolution that treats the above-mentioned concerns about contrast differently. The resolution predicted by this formula is proportional to the Rayleigh-based formula, differing by about 20%. For estimating theoretical resolution, it may be adequate.

$$r = \frac{\lambda}{2n\sin\theta} = \frac{\lambda}{2\text{NA}}$$

When a condenser is used to illuminate the sample, the shape of the pencil of light emanating from the condenser must also be included.

$$r = \frac{1.22\lambda}{\text{NA}_{obj} + \text{NA}_{cond}}$$

In a properly configured microscope, $\text{NA}_{obj} + \text{NA}_{cond} = 2\text{NA}_{obj}$.

The above estimates of resolution are specific to the case in which two identical very small samples that radiate incoherently in all directions. Other considerations must be taken into account if the sources radiate at different levels of intensity, are coherent, large, or radiate in non-uniform patterns.

Lens Resolution

The ability of a lens to resolve detail is usually determined by the quality of the lens but is ultimately limited by diffraction. Light coming from a point in the object diffracts through the lens aperture such that it forms a diffraction pattern in the image which has a central spot and surrounding bright rings, separated by dark nulls; this pattern is known as an Airy pattern, and the central bright lobe as an Airy disk. The angular radius of the Airy disk (measured from the center to the first null) is given by:

where

$$\theta = 1.22\frac{\lambda}{D}$$

θ is the angular resolution in radians,

λ is the wavelength of light in meters,

and D is the diameter of the lens aperture in meters.

Two adjacent points in the object give rise to two diffraction patterns. If the angular separation of the two points is significantly less than the Airy disk angular radius, then the two points cannot be resolved in the image, but if their angular separation is much greater than this, distinct images of the two points are formed and they can therefore be resolved. Rayleigh defined the somewhat arbitrary "Rayleigh criterion" that two points whose angular separation is equal to the Airy disk radius to first null can be considered to be resolved. It can be seen that the greater the diameter of the lens or its aperture, the greater the resolution. Astronomical telescopes have increasingly large lenses so they can 'see' ever finer detail in the stars.

Only the very highest quality lenses have diffraction limited resolution, however, and normally the quality of the lens limits its ability to resolve detail. This ability is expressed by the Optical Transfer Function which describes the spatial (angular) variation of the light signal as a function of spatial (angular) frequency. When the image is projected onto a flat plane, such as photographic film or a solid state detector, spatial frequency is the preferred domain, but when the image is referred to the lens alone, angular frequency is preferred. OTF may be broken down into the magnitude and phase components as follows:

$$OTF(\xi,\eta) = MTF(\xi,\eta) \cdot PTF(\xi,\eta)$$

where

$$MTF(\xi,\eta) = |OTF(\xi,\eta)|$$

$$PTF(\xi,\eta) = e^{-i2\cdot\pi\cdot\lambda(\xi,\eta)}$$

and (ξ,η) are spatial frequency in the x- and y-plane, respectively.

The OTF accounts for aberration, which the limiting frequency expression above does not. The magnitude is known as the Modulation Transfer Function (MTF) and the phase portion is known as the Phase Transfer Function (PTF).

In imaging systems, the phase component is typically not captured by the sensor. Thus, the important measure with respect to imaging systems is the MTF.

Phase is critically important to adaptive optics and holographic systems.

Sensor Resolution (Spatial)

Some optical sensors are designed to detect spatial differences in electromagnetic energy. These include photographic film, solid-state devices (CCD, CMOS detectors, and infrared detectors like PtSi and InSb), tube detectors (vidicon, plumbicon, and photomultiplier tubes used in night-vision devices), scanning detectors (mainly used for IR), pyroelectric detectors, and microbolometer detectors. The ability of such a detector to resolve those differences depends mostly on the size of the detecting elements.

Spatial resolution is typically expressed in line pairs per millimeter (lppmm), lines (of resolution, mostly for analog video), contrast vs. cycles/mm, or MTF (the modulus of OTF). The MTF may be found by taking the two-dimensional Fourier transform of the spatial sampling function. Smaller pixels result in wider MTF curves and thus better detection of higher frequency energy.

This is analogous to taking the Fourier transform of a signal sampling function; as in that case, the dominant factor is the sampling period, which is analogous to the size of the picture element (pixel).

Other factors include pixel noise, pixel cross-talk, substrate penetration, and fill factor.

A common problem among non-technicians is the use of the number of pixels on the detector to describe the resolution. If all sensors were the same size, this would be acceptable. Since they are not, the use of the number of pixels can be misleading. For example, a 2-megapixel camera of 20-micrometre-square pixels will have worse resolution than a 1-megapixel camera with 8-micrometre pixels, all else being equal.

For resolution measurement, film manufacturers typically publish a plot of Response (%) vs. Spatial Frequency (cycles per millimeter). The plot is derived experimentally. Solid state sensor and camera manufacturers normally publish specifications from which the user may derive a theoretical MTF according to the procedure outlined below. A few may also publish MTF curves, while others (especially intensifier manufacturers) will publish the response (%) at the Nyquist frequency, or, alternatively, publish the frequency at which the response is 50%.

To find a theoretical MTF curve for a sensor, it is necessary to know three characteristics of the sensor: the active sensing area, the area comprising the sensing area and the interconnection and support structures ("real estate"), and the total number of those areas (the pixel count). The total pixel count is almost always given. Sometimes the overall sensor dimensions are given, from which the real estate area can be calculated. Whether the real estate area is given or derived, if the active pixel area is not given, it may be derived from the real estate area and the fill factor, where fill factor is the ratio of the active area to the dedicated real estate area.

$$FF = \frac{a \cdot b}{c \cdot d}$$

where

- the active area of the pixel has dimensions $a \times b$

- the pixel real estate has dimensions $c \times d$

In Gaskill's notation, the sensing area is a 2D comb(x, y) function of the distance between pixels (the *pitch*), convolved with a 2D rect(x, y) function of the active area of the pixel, bounded by a 2D rect(x, y) function of the overall sensor dimension. The Fourier transform of this is a comb(ξ, η) function governed by the distance between pixels, convolved with a sinc(ξ, η) function governed by the number of pixels, and multiplied by the sinc(ξ, η) function corresponding to the active area. That last function serves as an overall envelope to the MTF function; so long as the number of pixels is much greater than one (1), then the active area size dominates the MTF.

Sampling function:

$$\mathbf{S}(x, y) = \left[\text{comb}\left(\frac{x}{c}, \frac{y}{d} \right) * \text{rect}\left(\frac{x}{a}, \frac{y}{b} \right) \right] \cdot \text{rect}\left(\frac{x}{M \cdot c}, \frac{y}{N \cdot d} \right)$$

where the sensor has $M \times N$ pixels

$$\mathbf{MTF}_{\text{sensor}}(\xi, \eta) = \mathcal{FF}(\mathbf{S}(x, y))$$

$$= [\text{sinc}((M \cdot c) \cdot \xi, (N \cdot d) \cdot \eta) * \text{comb}(c \cdot \xi, d \cdot \eta)] \cdot \text{sinc}(a \cdot \xi, b \cdot \eta)$$

Sensor Resolution (Temporal)

An imaging system running at 24 frames per second is essentially a discrete sampling system that samples a 2D area. The same limitations described by Nyquist apply to this system as to any signal sampling system.

All sensors have a characteristic time response. Film is limited at both the short resolution and the long resolution extremes by reciprocity breakdown. These are typically held to be anything longer than 1 second and shorter than 1/10,000 second. Furthermore, film requires a mechanical system to advance it through the exposure mechanism, or a moving optical system to expose it. These limit the speed at which successive frames may be exposed.

CCD and CMOS are the modern preferences for video sensors. CCD is speed-limited by the rate at which the charge can be moved from one site to another. CMOS has the advantage of having individually addressable cells, and this has led to its advantage in the high speed photography industry.

Vidicons, Plumbicons, and image intensifiers have specific applications. The speed at which they can be sampled depends upon the decay rate of the phosphor used. For example, the P46 phosphor has a decay time of less than 2 microseconds, while the P43 decay time is on the order of 2-3 milliseconds. The P43 is therefore unusable at frame rates above 1000 frames per second (frame/s).

Pyroelectric detectors respond to changes in temperature. Therefore, a static scene will not be detected, so they require choppers. They also have a decay time, so the pyroelectric system temporal response will be a bandpass, while the other detectors discussed will be a lowpass.

If objects within the scene are in motion relative to the imaging system, the resulting motion blur will result in lower spatial resolution. Short integration times will minimize the blur, but integration times are limited by sensor sensitivity. Furthermore, motion between frames in motion pictures will impact digital movie compression schemes (e.g. MPEG-1, MPEG-2). Finally, there are sampling schemes that require real or apparent motion inside the camera (scanning mirrors, rolling shutters) that may result in incorrect rendering of image motion. Therefore, sensor sensitivity and other time-related factors will have a direct impact on spatial resolution.

Analog Bandwidth Effect on Resolution

The spatial resolution of digital systems (e.g. HDTV and VGA) are fixed independently of the analog bandwidth because each pixel is digitized, transmitted, and stored as a discrete value. Digital cameras, recorders, and displays must be selected so that the resolution is identical from camera to display. However, in analog systems, the resolution of the camera, recorder, cabling, amplifiers, transmitters, receivers, and display may all be independent and the overall system resolution is governed by the bandwidth of the lowest performing component.

In analog systems, each horizontal line is transmitted as a high-frequency analog signal. Each picture element (pixel) is therefore converted to an analog electrical value (voltage), and changes in values between pixels therefore become changes in voltage. The transmission standards require that the sampling be done in a fixed time (outlined below), so more pixels per line becomes a requirement for more voltage changes per unit time, i.e. higher frequency. Since such signals are typically band-limited by cables, amplifiers, recorders, transmitters, and receivers, the band-limitation on the analog signal acts as an effective low-pass filter on the spatial resolution. The difference in resolutions between VHS (240 discernible lines per scanline), Betamax (280 lines), and the newer ED Beta format (500 lines) is explained primarily by the difference in the recording bandwidth.

In the NTSC transmission standard, each field contains 262.5 lines, and 59.94 fields are transmitted every second. Each line must therefore take 63 microseconds, 10.7 of which are for reset to the next line. Thus, the retrace rate is 15.734 kHz. For the picture to appear to have approximately the same horizontal and vertical resolution, it should be able to display 228 cycles per line, requiring a bandwidth of 4.28 MHz. If the line (sensor) width is known, this may be converted directly into cycles per millimeter, the unit of spatial resolution.

B/G/I/K television system signals (usually used with PAL colour encoding) transmit frames less often (50 Hz), but the frame contains more lines and is wider, so bandwidth requirements are similar.

Note that a "discernible line" forms one half of a cycle (a cycle requires a dark and a light line), so "228 cycles" and "456 lines" are equivalent measures.

System Resolution

There are two methods by which to determine system resolution. The first is to perform a series of two dimensional convolutions, first with the image and the lens, then the result of that procedure with the sensor, and so on through all of the components of the system. This is computationally expensive, and must be performed anew for each object to be imaged.

$$\textbf{Image}(x,y) = \ \textbf{Object}(x,y) * \textbf{PSF}_{\text{atmosphere}}(x,y) *$$

$$\textbf{PSF}_{\text{lens}}(x,y) * \textbf{PSF}_{\text{sensor}}(x,y) *$$

$$\textbf{PSF}_{\text{transmission}}(x,y) * \textbf{PSF}_{\text{display}}(x,y)$$

The other method is to transform each of the components of the system into the spatial frequency domain, and then to multiply the 2-D results. A system response may be determined without reference to an object. Although this method is considerably more difficult to comprehend conceptually, it becomes easier to use computationally, especially when different design iterations or imaged objects are to be tested.

The transformation to be used is the Fourier transform.

$$MTF_{\text{sys}}(\xi,\eta) = MTF_{atmosphere}(\xi,\eta).MTF_{lens}(\xi,\eta).$$

$$MTF_{sensor}(\xi,\eta).MTF_{transmission}(\xi,\eta)$$

$$MTF_{display}(\xi,\eta)$$

Ocular Resolution

The human eye is a limiting feature of many systems, when the goal of the system is to present data to humans for processing.

For example, in a security or air traffic control function, the display and work station must be constructed so that average humans can detect problems and direct corrective measures. Other examples are when a human is using eyes to carry out a critical task such as flying (piloting by visual reference), driving a vehicle, and so forth.

The best visual acuity of the human eye at its optical centre (the fovea) is less than 1 arc minute per line pair, reducing rapidly away from the fovea.

The human brain requires more than just a line pair to understand what the eye is imaging. Johnson's criteria defines the number of line pairs of ocular resolution, or sensor resolution, needed to recognize or identify an item.

Atmospheric Resolution

Systems looking through long atmospheric paths may be limited by turbulence. A key measure of the quality of atmospheric turbulence is the seeing diameter, also known as Fried's seeing diameter. A path which is temporally coherent is known as an *isoplanatic* patch.

Large apertures may suffer from aperture averaging, the result of several paths being integrated into one image.

Turbulence scales with wavelength at approximately a 6/5 power. Thus, seeing is better at infrared wavelengths than at visible wavelengths.

Short exposures suffer from turbulence less than longer exposures due to the "inner" and "outer" scale turbulence; short is considered to be much less than 10 ms for visible imaging (typically, anything less than 2 ms). Inner scale turbulence arises due to the eddies in the turbulent flow, while outer scale turbulence arises from large air mass flow. These masses typically move slowly, and so are reduced by decreasing the integration period.

A system limited only by the quality of the optics is said to be diffraction-limited. However, since atmospheric turbulence is normally the limiting factor for visible systems looking through long atmospheric paths, most systems are turbulence-limited. Corrections can be made by using adaptive optics or post-processing techniques.

$$\text{MTF}_s(v) = e^{-3.44 \cdot (\lambda f v / r_0)^{5/3} \cdot [1 - b \cdot (\lambda f v / D)^{1/3}]}$$

where

v is the spatial frequency

λ is the wavelength

f is the focal length

D is the aperture diameter

b is a constant (1 for far-field propagation)

and r_0 is Fried's seeing diameter

Measuring Optical Resolution

A variety of measurement systems are available, and use may depend upon the system being tested.

Typical test charts for Contrast Transfer Function (CTF) consist of repeated bar patterns. The limiting resolution is measured by determining the smallest group of bars, both vertically and horizontally, for which the correct number of bars can be seen. By calculating the contrast between the black and white areas at several different frequencies, however, points of the CTF can be determined with the contrast equation.

$$\text{contrast} = \frac{C_{max} - C_{min}}{C_{max} + C_{min}}$$

where

C_{max} is the normalized value of the maximum (for example, the voltage or grey value of the white area)

C_{min} is the normalized value of the minimum (for example, the voltage or grey value of the black area)

When the system can no longer resolve the bars, the black and white areas have the same value, so Contrast = 0. At very low spatial frequencies, C_{max} = 1 and C_{min} = 0 so Modulation = 1. Some modulation may be seen above the limiting resolution; these may be aliased and phase-reversed.

When using other methods, including the interferogram, sinusoid, and the edge in the

ISO 12233 target, it is possible to compute the entire MTF curve. The response to the edge is similar to a step response, and the Fourier Transform of the first difference of the step response yields the MTF.

Interferogram

An interferogram created between two coherent light sources may be used for at least two resolution-related purposes. The first is to determine the quality of a lens system, and the second is to project a pattern onto a sensor (especially photographic film) to measure resolution.

NBS 1010a/ ISO #2 Target

This 5 bar resolution test chart is often used for evaluation of microfilm systems and scanners. It is convenient for a 1:1 range (typically covering 1-18 cycles/mm) and is marked directly in cycles/mm. Details can be found in ISO-3334.

USAF 1951 Target

The USAF 1951 resolution test target consists of a pattern of 3 bar targets. Often found covering a range of 0.25 to 228 cycles/mm. Each group consists of six elements. The group is designated by a group number (-2, -1, 0, 1, 2, etc.) which is the power to which 2 should be raised to obtain the spatial frequency of the first element (e.g., group -2 is 0.25 line pairs per millimeter). Each element is the 6th root of 2 smaller than the preceding element in the group (e.g. element 1 is 2^0, element 2 is $2^{(-1/6)}$, element 3 is $2(-1/3)$, etc.). By reading off the group and element number of the first element which cannot be resolved, the limiting resolution may be determined by inspection. The complex numbering system and use of a look-up chart can be avoided by use of an improved but not standardized layout chart, which labels the bars and spaces directly in cycles/mm using OCR-A extended font.

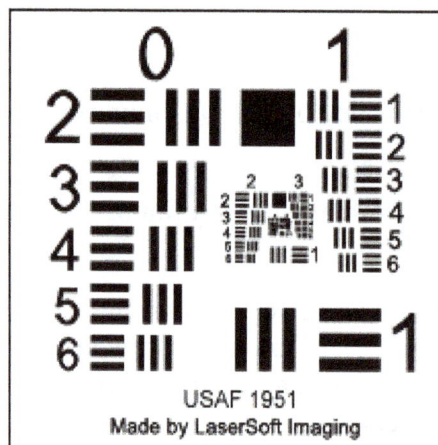

SilverFast Resolution Target USAF 1951 for determining a scanner's optimum resolution

$$Resolution = 2^{group+\frac{element-1}{6}}$$

NBS 1952 Target

The NBS 1952 target is a 3 bar pattern (long bars). The spatial frequency is printed alongside each triple bar set, so the limiting resolution may be determined by inspection. This frequency is normally only as marked after the chart has been reduced in size (typically 25 times). The original application called for placing the chart at a distance 26 times the focal length of the imaging lens used. The bars above and to the left are in sequence, separated by approximately the square root of two (12, 17, 24, etc.), while the bars below and to the left have the same separation but a different starting point (14, 20, 28, etc.)

EIA 1956 Video Resolution Target

The EIA 1956 resolution target was specifically designed to be used with television systems. The gradually expanding lines near the center are marked with periodic indications of the corresponding spatial frequency. The limiting resolution may be determined by inspection. The most important measure is the limiting horizontal resolution, since the vertical resolution is typically determined by the applicable video standard (I/B/G/K/NTSC/NTSC-J).

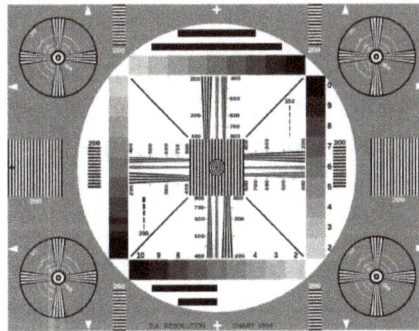

EIA 1956 video resolution target

IEEE Std 208-1995 Target

The IEEE 208-1995 resolution target is similar to the EIA target. Resolution is measured in horizontal and vertical TV lines.

ISO 12233 Target

The ISO 12233 target was developed for digital camera applications, since modern digital camera spatial resolution may exceed the limitations of the older targets. It includes several knife-edge targets for the purpose of computing MTF by Fourier transform.

They are offset from the vertical by 5 degrees so that the edges will be sampled in many different phases, which allow estimation of the spatial frequency response beyond the Nyquist frequency of the sampling.

Random Test Patterns

The idea is analogous to the use of a white noise pattern in acoustics to determine system frequency response.

Monotonically Increasing Sinusoid Patterns

The interferogram used to measure film resolution can be synthesized on personal computers and used to generate a pattern for measuring optical resolution.

Multiburst

A multiburst signal is an electronic waveform used to test analog transmission, recording, and display systems. The test pattern consists of several short periods of specific frequencies. The contrast of each may be measured by inspection and recorded, giving a plot of attenuation vs. frequency. The NTSC3.58 multiburst pattern consists of 500 kHz, 1 MHz, 2 MHz, 3 MHz, and 3.58 MHz blocks. 3.58 MHz is important because it is the chrominance frequency for NTSC video.

Discussion

It should be noted whenever using a bar target that the resulting measure is the contrast transfer function (CTF) and not the MTF. The difference arises from the subharmonics of the square waves and can be easily computed.

Interference (Wave Propagation)

The iridescence of soap bubbles is due to thin-film interference.

In physics, interference is a phenomenon in which two waves superpose to form a resultant wave of greater, lower, or the same amplitude. Interference usually refers to the interaction of waves that are correlated or coherent with each other, either because they come from the same source or because they have the same or nearly the same frequency. Interference effects can be observed with all types of waves, for example, light, radio, acoustic, surface water waves or matter waves.

Mechanism

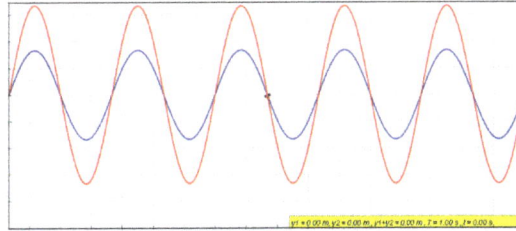

Interference of left traveling (green) and right traveling (blue) waves in one dimension, resulting in final (red) wave

Interference of waves from two point sources.

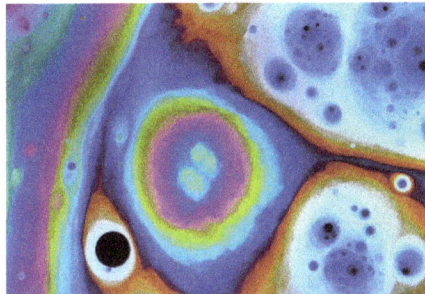

A magnified image of a coloured interference pattern in a soap film. The "black holes" are areas of almost total destructive interference, (antiphase).

The principle of superposition of waves states that when two or more propagating waves of same type are incident on the same point, the resultant amplitude at that point is equal to the vector sum of the amplitudes of the individual waves. If a crest of a wave meets a crest of another wave of the same frequency at the same point, then the amplitude is the sum of the individual amplitudes—this is constructive interference. If a crest of one wave meets a trough of another wave, then the amplitude is equal to the difference in the individual amplitudes—this is known as destructive interference.

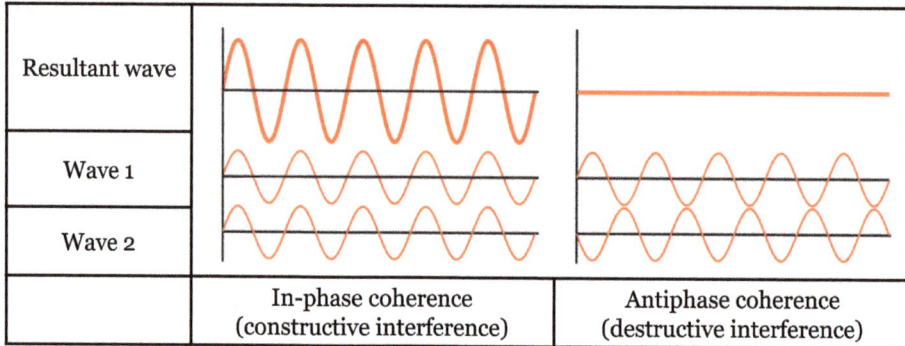

	In-phase coherence (constructive interference)	Antiphase coherence (destructive interference)
Resultant wave		
Wave 1		
Wave 2		

Constructive interference occurs when the phase difference between the waves is a multiple of 2π, whereas destructive interference occurs when the difference is an odd multiple of π. If the difference between the phases is intermediate between these two extremes, then the magnitude of the displacement of the summed waves lies between the minimum and maximum values.

Consider, for example, what happens when two identical stones are dropped into a still pool of water at different locations. Each stone generates a circular wave propagating outwards from the point where the stone was dropped. When the two waves overlap, the net displacement at a particular point is the sum of the displacements of the individual waves. At some points, these will be in phase, and will produce a maximum displacement. In other places, the waves will be in anti-phase, and there will be no net displacement at these points. Thus, parts of the surface will be stationary—these are seen in the figure above and to the right as stationary blue-green lines radiating from the center.

Between Two Plane Waves

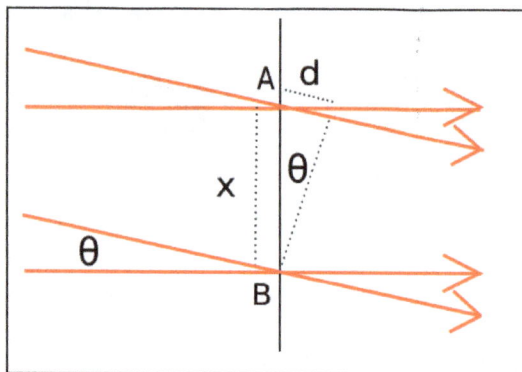

Geometrical arrangement for two plane wave interference

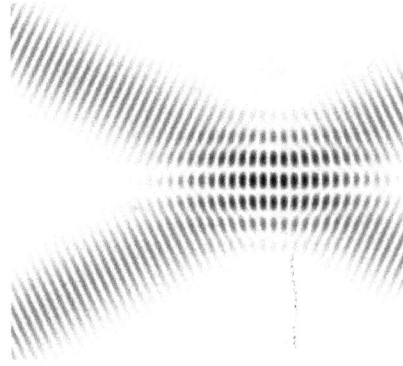

Interference fringes in overlapping plane waves

A simple form of interference pattern is obtained if two plane waves of the same frequency intersect at an angle. Interference is essentially an energy redistribution process. The energy which is lost at the destructive interference is regained at the constructive interference. One wave is travelling horizontally, and the other is travelling downwards at an angle θ to the first wave. Assuming that the two waves are in phase at the point B, then the relative phase changes along the x-axis. The phase difference at the point A is given by

$$\Delta\varphi = \frac{2\pi d}{\lambda} = \frac{2\pi x \sin\theta}{\lambda}$$

It can be seen that the two waves are in phase when

$$\frac{x\sin\theta}{\lambda} = 0, \pm 1, \pm 2, ...,$$

and are half a cycle out of phase when

$$\frac{x\sin\theta}{\lambda} = \pm\frac{1}{2}, \pm\frac{3}{2}, ...$$

Constructive interference occurs when the waves are in phase, and destructive interference when they are half a cycle out of phase. Thus, an interference fringe pattern is produced, where the separation of the maxima is

$$d_f = \frac{\lambda}{\sin\theta}$$

and d_f is known as the fringe spacing. The fringe spacing increases with increase in wavelength, and with decreasing angle θ.

The fringes are observed wherever the two waves overlap and the fringe spacing is uniform throughout.

Between Two Spherical Waves

Optical interference between two point sources for different wavelengths and source separations

A point source produces a spherical wave. If the light from two point sources overlaps, the interference pattern maps out the way in which the phase difference between the two waves varies in space. This depends on the wavelength and on the separation of the point sources. The figure to the right shows interference between two spherical waves. The wavelength increases from top to bottom, and the distance between the sources increases from left to right.

When the plane of observation is far enough away, the fringe pattern will be a series of almost straight lines, since the waves will then be almost planar.

Multiple Beams

Interference occurs when several waves are added together provided that the phase differences between them remain constant over the observation time.

It is sometimes desirable for several waves of the same frequency and amplitude to sum to zero (that is, interfere destructively, cancel). This is the principle behind, for example, 3-phase power and the diffraction grating. In both of these cases, the result is achieved by uniform spacing of the phases.

It is easy to see that a set of waves will cancel if they have the same amplitude and their phases are spaced equally in angle. Using phasors, each wave can be represented as $Ae^{i\varphi_n}$ for N waves from $n = 0$ to $n = N - 1$, where

$$\varphi_n - \varphi_{n-1} = \frac{2\pi}{N}.$$

To show that

$$\sum_{n=0}^{N-1} Ae^{i\varphi_n} = 0$$

one merely assumes the converse, then multiplies both sides by $e^{i\frac{2\pi}{N}}$

The Fabry–Pérot interferometer uses interference between multiple reflections.

A diffraction grating can be considered to be a multiple-beam interferometer, since the peaks which it produces are generated by interference between the light transmitted by each of the elements in the grating.

Optical Interference

Because the frequency of light waves (~10^{14} Hz) is too high to be detected by currently available detectors, it is possible to observe only the intensity of an optical interference pattern. The intensity of the light at a given point is proportional to the square of the average amplitude of the wave. This can be expressed mathematically as follows. The displacement of the two waves at a point r is:

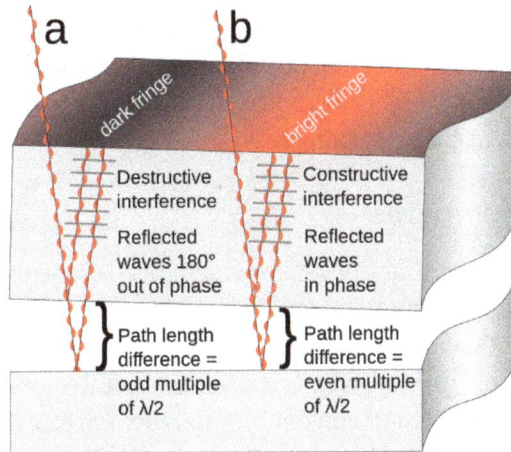

Creation of interference fringes by an optical flat on a reflective surface. Light rays from a monochromatic source pass through the glass and reflect off both the bottom surface of the flat and the supporting surface. The tiny gap between the surfaces means the two reflected rays have different path lengths and interfere when they combine. At locations (b) where the path difference is an even multiple of λ/2, the waves reinforce. At locations (a) where the path difference is an odd multiple of λ/2 the waves cancel. Since the gap between the surfaces varies slightly in width at different points, a series of alternating bright and dark bands are seen.

$$U_1(\mathbf{r},t) = A_1(\mathbf{r})e^{i[\varphi_1(\mathbf{r})-\omega t]}$$

$$U_2(\mathbf{r},t) = A_2(\mathbf{r})e^{i[\varphi_2(\mathbf{r})-\omega t]}$$

where A represents the magnitude of the displacement, φ represents the phase and ω represents the angular frequency.

The displacement of the summed waves is

$$U(\mathbf{r},t) = A_1(\mathbf{r})e^{i[\varphi_1(\mathbf{r})-\omega t]} + A_2(\mathbf{r})e^{i[\varphi_2(\mathbf{r})-\omega t]}$$

The intensity of the light at r is given by

$$I(\mathbf{r}) = \int U(\mathbf{r},t)U^*(\mathbf{r},t)dt \propto A_1^2(\mathbf{r}) + A_2^2(\mathbf{r}) + 2A_1(\mathbf{r})A_2(\mathbf{r})\cos[\varphi_1(\mathbf{r})-\varphi_2(\mathbf{r})]$$

This can be expressed in terms of the intensities of the individual waves as

$$I(\mathbf{r}) = I_1(\mathbf{r}) + I_2(\mathbf{r}) + 2\sqrt{I_1(\mathbf{r})I_2(\mathbf{r})}\cos[\varphi_1(\mathbf{r})-\varphi_2(\mathbf{r})]$$

Thus, the interference pattern maps out the difference in phase between the two waves, with maxima occurring when the phase difference is a multiple of 2π. If the two beams are of equal intensity, the maxima are four times as bright as the individual beams, and the minima have zero intensity.

The two waves must have the same polarization to give rise to interference fringes since it is not possible for waves of different polarizations to cancel one another out or add together. Instead, when waves of different polarization are added together, they give rise to a wave of a different polarization state.

Light Source Requirements

The discussion above assumes that the waves which interfere with one another are monochromatic, i.e. have a single frequency—this requires that they are infinite in time. This is not, however, either practical or necessary. Two identical waves of finite duration whose frequency is fixed over that period will give rise to an interference pattern while they overlap. Two identical waves which consist of a narrow spectrum of frequency waves of finite duration, will give a series of fringe patterns of slightly differing spacings, and provided the spread of spacings is significantly less than the average fringe spacing, a fringe pattern will again be observed during the time when the two waves overlap.

Conventional light sources emit waves of differing frequencies and at different times from different points in the source. If the light is split into two waves and then re-combined, each individual light wave may generate an interference pattern with its other half, but the individual fringe patterns generated will have different phases and spacings, and normally no overall fringe pattern will be observable. However, single-element light sources, such as sodium- or mercury-vapor lamps have emission lines with quite narrow frequency spectra. When these are spatially and colour filtered, and then split into two waves, they can be superimposed to generate interference fringes. All interferometry prior to the invention of the laser was done using such sources and had a wide range of successful applications.

A laser beam generally approximates much more closely to a monochromatic source, and it is much more straightforward to generate interference fringes using a laser. The ease with which interference fringes can be observed with a laser beam can sometimes cause problems in that stray reflections may give spurious interference fringes which can result in errors.

Normally, a single laser beam is used in interferometry, though interference has been observed using two independent lasers whose frequencies were sufficiently matched to satisfy the phase requirements.

It is also possible to observe interference fringes using white light. A white light fringe pattern can be considered to be made up of a 'spectrum' of fringe patterns each of slightly different spacing. If all the fringe patterns are in phase in the centre, then the fringes will increase in size as the wavelength decreases and the summed intensity will show three to four fringes of varying colour. Young describes this very elegantly in his discussion of two slit interference. Some fine examples of white light fringes can be seen here. Since white light fringes are obtained only when the two waves have travelled equal distances from the light source, they can be very useful in interferometry, as they allow the zero path difference fringe to be identified.

White light interference in a soap bubble

Optical Arrangements

To generate interference fringes, light from the source has to be divided into two waves which have then to be re-combined. Traditionally, interferometers have been classified as either amplitude-division or wavefront-division systems.

In an amplitude-division system, a beam splitter is used to divide the light into two beams travelling in different directions, which are then superimposed to produce the interference pattern. The Michelson interferometer and the Mach-Zehnder interferometer are examples of amplitude-division systems.

In wavefront-division systems, the wave is divided in space—examples are Young's double slit interferometer and Lloyd's mirror.

Interference can also be seen in everyday phenomena such as iridescence and structural coloration. For example, the colours seen in a soap bubble arise from interference of light reflecting off the front and back surfaces of the thin soap film. Depending on the thickness of the film, different colours interfere constructively and destructively.

Applications

Optical Interferometry

Interferometry has played an important role in the advancement of physics, and also has a wide range of applications in physical and engineering measurement.

Thomas Young's double slit interferometer in 1803 demonstrated interference fringes when two small holes were illuminated by light from another small hole which was illuminated by sunlight. Young was able to estimate the wavelength of different colours in the spectrum from the spacing of the fringes. The experiment played a major role in the general acceptance of the wave theory of light. In quantum mechanics, this experiment is considered to demonstrate the inseparability of the wave and particle natures of light and other quantum particles (wave–particle duality). Richard Feynman was fond of saying that all of quantum mechanics can be gleaned from carefully thinking through the implications of this single experiment.

The results of the Michelson–Morley experiment are generally considered to be the first strong evidence against the theory of a luminiferous aether and in favor of special relativity.

Interferometry has been used in defining and calibrating length standards. When the metre was defined as the distance between two marks on a platinum-iridium bar, Michelson and Benoît used interferometry to measure the wavelength of the red cadmium line in the new standard, and also showed that it could be used as a length standard. Sixty years later, in 1960, the metre in the new SI system was defined to be equal to 1,650,763.73 wavelengths of the orange-red emission line in the electromagnetic spectrum of the krypton-86 atom in a vacuum. This definition was replaced in 1983 by defining the metre as the distance travelled by light in vacuum during a specific time interval. Interferometry is still fundamental in establishing the calibration chain in length measurement.

Interferometry is used in the calibration of slip gauges (called gauge blocks in the US) and in coordinate-measuring machines. It is also used in the testing of optical components.

Radio Interferometry

In 1946, a technique called astronomical interferometry was developed. Astronomical radio interferometers usually consist either of arrays of parabolic dishes or two-dimensional arrays of omni-directional antennas. All of the telescopes in the array are widely separated and are usually connected together using coaxial cable, waveguide, optical fiber, or other type of transmission line. Interferometry increases the total signal collected, but its primary purpose is to vastly increase the resolution through a process called Aperture synthesis. This technique works by superposing (interfering) the signal waves from the different telescopes on the principle that waves that coincide with the same phase will add to each other while two waves that have opposite phases will cancel each other out. This creates a combined telescope that is equivalent in resolution (though not in sensitivity) to a single antenna whose diameter is equal to the spacing of the antennas furthest apart in the array.

The Very Large Array, an interferometric array formed from many smaller telescopes, like many larger radio telescopes.

Acoustic Interferometry

An acoustic interferometer is an instrument for measuring the physical characteristics of sound wave in a gas or liquid. It may be used to measure velocity, wavelength, absorption, or impedance. A vibrating crystal creates the ultrasonic waves that are radiated into the medium. The waves strike a reflector placed parallel to the crystal. The waves are then reflected back to the source and measured.

Quantum Interference

If a system is in state ψ, its wavefunction is described in Dirac or bra–ket notation as:

$$|\psi\rangle = \sum_i |i\rangle\langle i|\psi\rangle = \sum_i |i\rangle\psi_i$$

where the $|i\rangle$s specify the different quantum "alternatives" available (technically, they form an eigenvector basis) and the ψ_i are the probability amplitude coefficients, which are complex numbers.

The probability of observing the system making a transition or quantum leap from state ψ to a new state φ is the square of the modulus of the scalar or inner product of the two states:

$$\text{prob}(\psi \Rightarrow \varphi) = |\langle \psi | \varphi \rangle|^2 = \left| \sum_i \psi_i^* \varphi_i \right|^2$$

$$= \sum_{ij} \psi_i^* \psi_j \varphi_j^* \varphi_i = \sum_i |\psi_i|^2 |\varphi_i|^2 + \sum_{ij;i \neq j} \psi_i^* \psi_j \varphi_j^* \varphi_i$$

where $\psi_i = \langle i | \psi \rangle$ (as defined above) and similarly $\varphi_i = \langle i | \varphi \rangle$ are the coefficients of the final state of the system. * is the complex conjugate so that $\psi_i^* = \langle \psi | i \rangle$, etc.

Now let's consider the situation classically and imagine that the system transited from $| \psi \rangle$ to $| \varphi \rangle$ via an intermediate state $| i \rangle$. Then we would *classically* expect the probability of the two-step transition to be the sum of all the possible intermediate steps. So we would have

$$\text{prob}(\psi \Rightarrow \varphi) = \sum_i \text{prob}(\psi \Rightarrow i \Rightarrow \varphi)$$

$$= \sum_i |\langle \psi | i \rangle|^2 |\langle i | \varphi \rangle|^2 = \sum_i |\psi_i|^2 |\varphi_i|^2,$$

The classical and quantum derivations for the transition probability differ by the presence, in the quantum case, of the extra terms $\sum_{ij;i \neq j} \psi_i^* \psi_j \varphi_j^* \varphi_i$; these extra quantum terms represent *interference* between the different $i \neq j$ intermediate "alternatives". These are consequently known as the quantum interference terms, or cross terms. This is a purely quantum effect and is a consequence of the non-additivity of the probabilities of quantum alternatives.

The interference terms vanish, via the mechanism of quantum decoherence, if the intermediate state $| i \rangle$ is measured or coupled with the environment.

Optical Illusion

The checker shadow illusion. Although square A appears a darker shade of grey than square B, the two are exactly the same.

An optical illusion (also called a visual illusion) is an illusion caused by the eye and characterized by visually perceived images that differ from objective reality. The information gathered by the eye is processed in the brain to give a perception that does not tally with a physical measurement of the stimulus source. There are three main types: literal optical illusions that create images that are different from the objects that make them, physiological illusions that are the effects of excessive stimulation of a specific type (brightness, colour, size, position, tilt, movement), and cognitive illusions, the result of unconscious inferences. Pathological visual illusions arise from a pathological exaggeration in physiological visual perception mechanisms causing the aforementioned types of illusions.

Optical illusions are often classified into categories including the physical and the cognitive or perceptual, and contrasted with optical hallucinations.

In this animation, Mach bands exaggerate the contrast between edges of the slightly differing shades of gray, as soon as they come in contact with one-another.

Physiological illusions, such as the afterimages following bright lights, or adapting stimuli of excessively longer alternating patterns (contingent perceptual aftereffect), are presumed to be the effects on the eyes or brain of excessive stimulation or interaction with contextual or competing stimuli of a specific type—brightness, color, position, tile, size, movement, etc. The theory is that a stimulus follows its individual dedicated neural path in the early stages of visual processing, and that intense or repetitive activity in that or interaction with active adjoining channels cause a physiological imbalance that alters perception.

The Hermann grid illusion and Mach bands are two illusions that are best explained using a biological approach. Lateral inhibition, where in the receptive field of the retina light and dark receptors compete with one another to become active, has been used to explain why we see bands of increased brightness at the edge of a color difference when viewing Mach bands. Once a receptor is active, it inhibits adjacent receptors. This inhibition creates contrast, highlighting edges. In the Hermann grid illusion the gray spots appear at the intersection because of the inhibitory response which occurs as a result of the increased dark surround. Lateral inhibition has also been used to explain the Hermann grid illusion, but this has been disproved. More recent empirical approaches to optical illusions have had some success in explaining optical phenomena with which theories based on lateral inhibition have struggled (e.g. Howe et al. 2005).

Pathological Visual Illusions

A pathological visual illusion is a distortion of a real external stimulus and are often diffuse and persistent. Pathological visual illusions usually occur throughout the visual field, suggesting global excitability or sensitivity alterations. Alternatively visual hallucination is the perception of an external visual stimulus where none exists. Visual hallucinations are often from focal dysfunction and are usually transient.

Types of visual illusions include oscillopsia, halos around objects, illusory palinopsia (visual trailing, light streaking, prolonged indistinct afterimages), akinetopsia, visual snow, micropsia, macropsia, teleopsia, pelopsia, Alice in Wonderland syndrome, metamorphopsia, dyschromatopsia, intense glare, blue field entoptic phenomenon, and purkinje trees.

These symptoms may indicate an underlying disease state and necessitate seeing a medical practitioner. Etiologies associated with pathological visual illusions include multiple types of ocular disease, migraines, hallucinogen persisting perception disorder, head trauma, and prescription drugs. If a medical work-up does not reveal a cause of the pathological visual illusions, the idiopathic visual disturbances could be analogous to the altered excitability state seen in visual aura with no migraine headache. If the visual illusions are diffuse and persistent, they often affect the patient's quality of life. These symptoms are often refractory to treatment and may be caused by any of the aforementioned etiologes, but are often idiopathic. There is no standard treatment for these visual disturbances.

Cognitive Illusions

Cognitive illusions are assumed to arise by interaction with assumptions about the world, leading to "unconscious inferences", an idea first suggested in the 19th century by the German physicist and physician Hermann Helmholtz. Cognitive illusions are commonly divided into ambiguous illusions, distorting illusions, paradox illusions, or fiction illusions.

1. Ambiguous illusions are pictures or objects that elicit a perceptual "switch" between the alternative interpretations. The Necker cube is a well-known example; another instance is the Rubin vase.

2. Distorting or geometrical-optical illusions are characterized by distortions

of size, length, position or curvature. A striking example is the Café wall illusion. Other examples are the famous Müller-Lyer illusion and Ponzo illusion.

3. Paradox illusions are generated by objects that are paradoxical or impossible, such as the Penrose triangle or impossible staircase seen, for example, in M.C. Escher's *Ascending and Descending* and *Waterfall*. The triangle is an illusion dependent on a cognitive misunderstanding that adjacent edges must join.

4. Fictions are when a figure is perceived even though it is not in the stimulus.

Explanation of Cognitive Illusions

Perceptual Organization

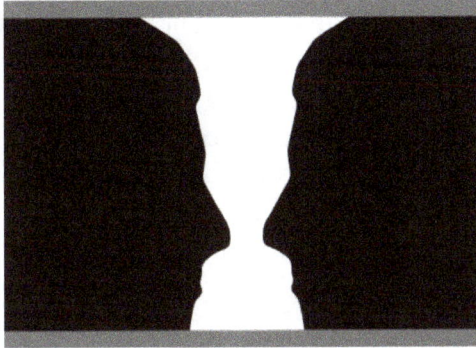

Reversible figures and vase, or the figure-ground illusion

Rabbit–duck illusion

To make sense of the world it is necessary to organize incoming sensations into information which is meaningful. Gestalt psychologists believe one way this is done is by perceiving individual sensory stimuli as a meaningful whole. Gestalt organization can be used to explain many illusions including the rabbit–duck illusion where the image as a whole switches back and forth from being a duck then being a rabbit and why in the figure–ground illusion the figure and ground are reversible.

Kanizsa's Triangle

In addition, Gestalt theory can be used to explain the illusory contours in the Kaniz-sa's Triangle. A floating white triangle, which does not exist, is seen. The brain has a need to see familiar simple objects and has a tendency to create a "whole" image from individual elements. Gestalt means "form" or "shape" in German. However, another explanation of the Kanizsa's Triangle is based in evolutionary psychology and the fact that in order to survive it was important to see form and edges. The use of perceptual organization to create meaning out of stimuli is the principle behind other well-known illusions including impossible objects. Our brain makes sense of shapes and symbols putting them together like a jigsaw puzzle, formulating that which isn't there to that which is believable.

The Gestalt principles of perception govern the way we group different objects. Good form is where the perceptual system tries to fill in the blanks in order to see simple objects rather than complex objects. Continuity is where the perceptual system tries to disambiguate which segments fit together into continuous lines. Proximity is where objects that are close together are associated. Similarity is where objects that are similar are seen as associated. Some of these elements have been successfully incorporated into quantitative models involving optimal estimation or Bayesian inference.

The double-anchoring theory, a popular but recent theory of cognitive illusions, states that any region belongs to a framework of Gestalt principles and within each framework, is independently anchored a highest surrounding brightness and highest brightness. A spot's lightness is determined by using the average of values of brightness in each framework.

Depth and Motion Perception

Illusions can be based on an individual's ability to see in three dimensions even though the image hitting the retina is only two dimensional. The Ponzo illusion is an example of an illusion which uses monocular cues of depth perception to fool the eye. But even with two dimensional images, the brain exaggerates vertical distances when compared with horizontal distances, as in the vertical-horizontal illusion where the two lines are exactly the same length.

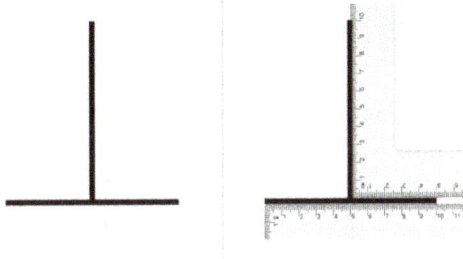

The vertical–horizontal illusion where the vertical line is thought to be longer than the horizontal

In the Ponzo illusion the converging parallel lines tell the brain that the image higher in the visual field is farther away therefore the brain perceives the image to be larger, although the two images hitting the retina are the same size. The optical illusion seen in a diorama/false perspective also exploits assumptions based on monocular cues of depth perception. The M.C. Escher painting *Waterfall* exploits rules of depth and proximity and our understanding of the physical world to create an illusion. Like depth perception, motion perception is responsible for a number of sensory illusions. Film animation is based on the illusion that the brain perceives a series of slightly varied images produced in rapid succession as a moving picture. Likewise, when we are moving, as we would be while riding in a vehicle, stable surrounding objects may appear to move. We may also perceive a large object, like an airplane, to move more slowly than smaller objects, like a car, although the larger object is actually moving faster. The phi phenomenon is yet another example of how the brain perceives motion, which is most often created by blinking lights in close succession.

The ambiguity of direction of motion due to lack of visual references for depth is shown in the spinning dancer illusion. The spinning dancer appears to be moving clockwise or counterclockwise depending on spontaneous activity in the brain where perception is subjective. Recent studies show on the fMRI that there are spontaneous fluctuations in cortical activity while watching this illusion, particularly the parietal lobe, because it is involved in perceiving movement.

Color and Brightness Constancies

Perceptual constancies are sources of illusions. Color constancy and brightness constancy are responsible for the fact that a familiar object will appear the same color regardless of the amount of light or color of light reflecting from it. An illusion of color or contrast difference can be created when the luminosity or color of the area surrounding an unfamiliar object is changed. The contrast of the object will appear darker against a black field that reflects less light compared to a white field even though the object itself did not change in color. Similarly, the eye will compensate for color contrast depending on the color cast of the surrounding area.

Simultaneous Contrast Illusion. The background is a color gradient and progresses from dark grey to light grey. The horizontal bar appears to progress from light grey to dark grey, but is in fact just one colour.

In addition to the Gestalt principles of perception, water-color illusions contribute to the formation of optical illusions. Water-color illusions consist of object-hole effects and coloration. Object-hole effects occur when boundaries are prominent where there is a figure and background with a hole that is 3D volumetric in appearance. Coloration consists of an assimilation of color radiating from a thin-colored edge lining a darker chromatic contour. The water-color illusion describes how the human mind perceives the wholeness of an object such as top-down processing. Thus, contextual factors play into perceiving the brightness of an object.

Object

Just as it perceives color and brightness constancies, the brain has the ability to understand familiar objects as having a consistent shape or size. For example, a door is perceived as rectangle regardless of how the image may change on the retina as the door is opened and closed. Unfamiliar objects, however, do not always follow the rules of shape constancy and may change when the perspective is changed. The Shepard illusion of the changing table is an example of an illusion based on distortions in shape constancy.

Future Perception

Researcher Mark Changizi of Rensselaer Polytechnic Institute in New York has a more imaginative take on optical illusions, saying that they are due to a neural lag which most humans experience while awake. When light hits the retina, about one-tenth of a second goes by before the brain translates the signal into a visual perception of the world. Scientists have known of the lag, yet they have debated how humans compensate, with some proposing that our motor system somehow modifies our movements to offset the delay.

Changizi asserts that the human visual system has evolved to compensate for neural delays by generating images of what will occur one-tenth of a second into the future. This foresight enables humans to react to events in the present, enabling humans to perform reflexive acts like catching a fly ball and to maneuver smoothly through a crowd.

Illusions occur when our brains attempt to perceive the future, and those perceptions don't match reality. For example, an illusion called the Hering illusion looks like bicycle spokes around a central point, with vertical lines on either side of this central, so-called vanishing point. The illusion tricks us into thinking we are moving forward, and thus, switches on our future-seeing abilities. Since we aren't actually moving and the figure is static, we misperceive the straight lines as curved ones.

Changizi said:

Evolution has seen to it that geometric drawings like this elicit in us premonitions of the near future. The converging lines toward a vanishing point (the spokes) are cues that trick our brains into thinking we are moving forward—as we would in the real world, where the door frame (a pair of vertical lines) seems to bow out as we move through it—and we try to perceive what that world will look like in the next instant.

Illusions

In Art

Artists who have worked with optical illusions include M. C. Escher, Bridget Riley, Salvador Dalí, Giuseppe Arcimboldo, Patrick Bokanowski, Marcel Duchamp, Jasper Johns, Oscar Reutersvärd, Victor Vasarely and Charles Allan Gilbert. Contemporary artists who have experimented with illusions include Jonty Hurwitz, Sandro del Prete, Octavio Ocampo, Dick Termes, Shigeo Fukuda, Patrick Hughes, István Orosz, Rob Gonsalves, Gianni A. Sarcone, Ben Heine and Akiyoshi Kitaoka. Optical illusion is also used in film by the technique of forced perspective.

Op art is a style of art that uses optical illusions to create an impression of movement, or hidden images and patterns. *Trompe-l'œil* uses realistic imagery to create the optical illusion that depicted objects exist in three dimensions.

Cognitive Processes Hypothesis

The hypothesis claims that visual illusions occur because the neural circuitry in our visual system evolves, by neural learning, to a system that makes very efficient interpretations of usual 3D scenes based in the emergence of simplified models in our brain that speed up the interpretation process but give rise to optical illusions in unusual situations. In this sense, the cognitive processes hypothesis can be considered a framework for an understanding of optical illusions as the signature of the empirical statistical way vision has evolved to solve the inverse problem.

Research indicates that 3D vision capabilities emerge and are learned jointly with the planning of movements. After a long process of learning, an internal representation of the world emerges that is well-adjusted to the perceived data coming from closer objects. The representation of distant objects near the horizon is less "adequate". In fact, it is not only the Moon that seems larger when we perceive it near the horizon. In a photo of a distant scene, all distant objects are perceived as smaller than when we observe them directly using our vision.

The retinal image is the main source driving vision but what we see is a "virtual" 3D representation of the scene in front of us. We don't see a physical image of the world; we see objects, and the physical world is not itself separated into objects. We see it according to the way our brain organizes it. The names, colours, usual shapes and other information about the things we see pop up instantaneously from our neural circuitry and influence the representation of the scene. We "see" the most relevant information about the elements of the best 3D image that our neural networks can produce. The illusions arise when the "judgments" implied in the unconscious analysis of the scene are in conflict with reasoned considerations about it.

Birefringence

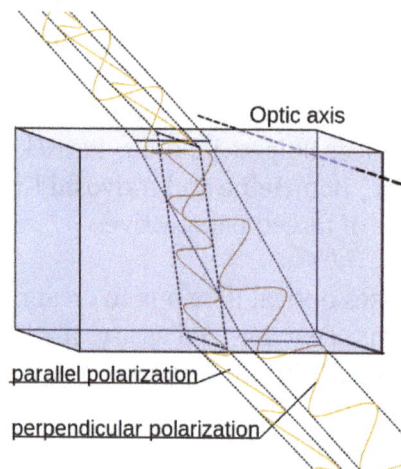

Incoming light in the parallel (s) polarization sees a different effective index of refraction than light in the perpendicular (p) polarization, and is thus refracted at a different angle.

A calcite crystal laid upon a graph paper with blue lines showing the double refraction

Doubly refracted image as seen through a calcite crystal, seen through a rotating polarizing filter illustrating the opposite polarization states of the two images.

Birefringence is the optical property of a material having a refractive index that depends on the polarization and propagation direction of light. These optically anisotropic materials are said to be birefringent (or birefractive). The birefringence is often quantified as the maximum difference between refractive indices exhibited by the material. Crystals with non-cubic crystal structures are often birefringent, as are plastics under mechanical stress.

Birefringence is responsible for the phenomenon of double refraction whereby a ray of light, when incident upon a birefringent material, is split by polarization into two rays taking slightly different paths. This effect was first described by the Danish scientist Rasmus Bartholin in 1669, who observed it in calcite, a crystal having one of the strongest birefringences. However it was not until the 19th century that Augustin-Jean Fresnel described the phenomenon in terms of polarization, understanding light as a wave with field components in transverse polarizations (perpendicular to the direction of the wave vector).

Explanation

A mathematical description of wave propagation in a birefringent medium is presented below. Following is a qualitative explanation of the phenomenon.

Uniaxial Materials

The simplest (and most common) type of birefringence is described as *uniaxial*, meaning that there is a single direction governing the optical anisotropy whereas all directions perpendicular to it (or at a given angle to it) are optically equivalent. Thus rotating the material around this axis does not change its optical behavior. This special direction is known as the optic axis of the material. Light whose polarization is *perpendicular* to the optic axis is governed by a refractive index n_o (for "ordinary"). Light whose polarization is *in the direction of* the optic axis sees an optical index n_e (for "extraordinary"). For any ray direction there is a linear polarization direction perpendicular to the optic axis, and this is called an *ordinary ray*. However, for ray directions not parallel to the optic axis, the polarization direction perpendicular to the ordinary ray's polarization will be partly in the direction of the optic axis, and this is called an *extraordinary ray*. The ordinary ray will always experience a refractive index of n_o, whereas the refractive index of the extraordinary ray will be in between n_o and n_e, depending on the ray direction as described by the index ellipsoid. The magnitude of the difference is quantified by the birefringence:

$$\Delta n = n_e - n_o..$$

The propagation (as well as reflection coefficient) of the ordinary ray is simply described by n_o as if there were no birefringence involved. However the extraordinary ray, as its name suggests, propagates unlike any wave in a homogenous optical material. Its refraction (and reflection) at a surface can be understood using the effective refractive index (a value in between n_o and n_e). However it is in fact an inhomogeneous wave whose power flow (given by the Poynting vector) is not exactly in the direction of the wave vector. This causes an additional shift in that beam, even when launched at normal incidence, as is popularly observed using a crystal of calcite as photographed above. Rotating the calcite crystal will cause one of the two images, that of the extraordinary ray, to rotate slightly around that of the ordinary ray, which remains fixed.

When the light propagates either along or orthogonal to the optic axis, such a lateral shift does not occur. In the first case, both polarizations see the same effective refractive index, so there is no extraordinary ray. In the second case the extraordinary ray propagates at a different phase velocity (corresponding to n_e) but is not an inhomogeneous wave. A crystal with its optic axis in this orientation, parallel to the optical surface, may be used to create a waveplate, in which there is no distortion of the image but an intentional modification of the state of polarization of the incident wave. For instance, a quarter-wave plate is commonly used to create circular polarization from a linearly polarized source.

Biaxial Materials

The case of so-called biaxial crystals is substantially more complex. These are characterized by *three* refractive indices corresponding to three principal axes of the crystal. For most ray directions, *both* polarizations would be classified as extraordinary rays but with different effective refractive indices. Being extraordinary waves, however, the direction of power flow is not identical to the direction of the wave vector in either case.

The two refractive indices can be determined using the index ellipsoids for given directions of the polarization. Note that for biaxial crystals the index ellipsoid will *not* be an ellipsoid of revolution ("spheroid") but is described by three unequal principle refractive indices n_α, n_β and n_γ. Thus there is no axis around which a rotation leaves the optical properties invariant (as there is with uniaxial crystals whose index ellipsoid *is* a spheroid).

Although there is no axis of symmetry, there are *two* optical axes or *binormals* which are defined as directions along which light may propagate without birefringence, i.e., directions along which the wavelength is independent of polarization. For this reason, birefringent materials with three distinct refractive indices are called *biaxial*. Additionally, there are two distinct axes known as *optical ray axes* or *biradials* along which the group velocity of the light is independent of polarization.

Double Refraction

When an arbitrary beam of light strikes the surface of a birefringent material, the polarizations corresponding to the ordinary and extraordinary rays generally take somewhat different paths. Unpolarized light consists of equal amounts of energy in any two orthogonal polarizations, and even polarized light (except in special cases) will have some energy in each of these polarizations. According to Snell's law of refraction, the angle of refraction will be governed by the effective refractive index which is different between these two polarizations. This is clearly seen, for instance, in the Wollaston prism which is designed to separate incoming light into two linear polarizations using a birefringent material such as calcite.

The different angles of refraction for the two polarization components are shown in the figure at the top of the page, with the optic axis along the surface (and perpendicular to the plane of incidence), so that the angle of refraction is different for the *p* polarization (the "ordinary ray" in this case, having its polarization perpendicular to the optic axis) and the *s* polarization (the "extraordinary ray" with a polarization component along the optic axis). In addition, a distinct form of double refraction occurs in cases where the optic axis is not along the refracting surface (nor exactly normal to it); in this case the electric polarization of the birefringent material is not exactly in the direction of the wave's electric field for the extraordinary ray. The direction of power flow (given by the Poynting vector) for this inhomogenous wave is at a finite angle from the direction of the wave vector resulting in an additional separation between these beams. So even in the case of normal incidence, where the angle of refraction is zero (according to Snell's law, regardless of effective index of refraction), the energy of the extraordinary ray may be propagated at an angle. This is commonly observed using a piece of calcite cut appropriately with respect to its optic axis, placed above a paper with writing, as in the above two photographs.

Terminology

Much of the work involving polarization preceded the understanding of light as a transverse electromagnetic wave, and this has affected some terminology in use. Isotropic materials have symmetry in all directions and the refractive index is the same for any polarization direction. An anisotropic material is called "birefringent" because it will generally refract a single incoming ray in two directions, which we now understand correspond to the two different polarizations. This is true of either a uniaxial or biaxial material.

In a uniaxial material, one ray behaves according to the normal law of refraction (corresponding to the ordinary refractive index), so an incoming ray at normal incidence remains normal to the refracting surface. However, as explained above, the other polarization can be deviated from normal incidence, which cannot be described using the law of refraction. This thus became known as the *extraordinary ray*. The terms "ordinary" and "extraordinary" are still applied to the polarization components perpendicular to and not perpendicular to the optic axis respectively, even in cases where no double refraction is involved.

A material is termed *uniaxial* when it has a single direction of symmetry in its optical behavior, which we term the optic axis. It also happens to be the axis of symmetry of the index ellipsoid (a spheroid in this case). The index ellipsoid could still be described according to the refractive indices, n_α, n_β and n_γ, along three coordinate axes, however in this case two are equal. So if $n_\alpha = n_\beta$ corresponding to the x and y axes, then the extraordinary index is n_γ corresponding to the z axis, which is also called the *optic axis* in this case.

However materials in which all three refractive indices are different are termed *biaxial* and the origin of this term is more complicated and frequently misunderstood. In a uniaxial crystal, different polarization components of a beam will travel at different phase velocities, *except* for rays in the direction of what we call the optic axis. Thus the optic axis has the particular property that rays in that direction do *not* exhibit birefringence, with all polarizations in such a beam experiencing the same index of refraction. It is very different when the three principal refractive indices are all different; then an incoming ray in any of those principle directions will still encounter two different refractive indices. But it turns out that there are two special directions (at an angle to all of the 3 axes) where the refractive indices for different polarizations are again equal. For this reason, these crystals were designated as *biaxial*, with the two "axes" in this case referring to ray directions in which propagation does not experience birefringence.

Fast and Slow Rays

In a birefringent material, a wave consists of two polarization components which generally are governed by different effective refractive indices. The so-called *slow ray* is the component for which the material has the higher effective refractive index (slower phase velocity), while the *fast ray* is the one with a lower effective refractive index. When a beam is incident on such a material from air (or any material with a lower refractive index), the slow ray is thus refracted more towards the normal than the fast ray. In the figure at the top of the page, it can be seen that refracted ray with s polarization in the direction of the optic axis (thus the extraordinary ray) is the slow ray in this case.

Using a thin slab of that material at normal incidence, one would implement a waveplate. In this case there is essentially no spatial separation between the polarizations, however the phase of the wave in the parallel polarization (the slow ray) will be retarded with respect to the perpendicular polarization. These directions are thus known as the slow axis and fast axis of the waveplate.

Positive or Negative

Uniaxial birefringence is classified as positive when the extraordinary index of refraction n_e is greater than the ordinary index n_o. Negative birefringence means that $\Delta n = n_e - n_o$ is less than zero. In other words, the polarization of the fast (or slow) wave is perpendicular to the optic axis when the birefringence of the crystal is positive (or negative, respectively). In the case of biaxial crystals, all three of the principal axes have different refractive indices so this designation does not apply. But for any defined ray direction one can just as well designate the fast and slow ray polarizations.

Sources of Optical Birefringence

While birefringence is usually obtained using an anisotropic crystal, it can result from an optically isotropic material in a few ways:

- Stress birefringence results when isotropic materials are stressed or deformed (i.e., stretched or bent) causing a loss of physical isotropy and consequently a loss of isotropy in the material's permittivity tensor.

- Circular birefringence in liquids where there is a enantiomeric excess in a solution containing a molecule which has stereo isomers.

- Form birefringence, whereby structure elements such as rods, having one refractive index, are suspended in a medium with a different refractive index. When the lattice spacing is much smaller than a wavelength, such a structure is described as a metamaterial.

- By the Kerr effect, whereby an applied electric field induces birefringence at optical frequencies through the effect of nonlinear optics;

- By the Faraday effect, where a magnetic field causes some materials to become circularly birefringent (having slightly different indices of refraction for left and right handed circular polarizations), making the material optically active until the field is removed;

- By the self or forced alignment into thin films of amphiphilic molecules such as lipids, some surfactants or liquid crystals

Common Birefringent Materials

The best-characterized birefringent materials are crystals. Due to their specific crystal structures their refractive indices are well defined. Depending on the symmetry of a crystal structure (as determined by one of the 32 possible crystallographic point groups), crystals in that group may be forced to be isotropic (not birefringent), to have uniaxial symmetry, or neither in which case it is a biaxial crystal. The crystal structures permitting uniaxial and biaxial birefringence are noted in the two tables, below, listing the two or three principal refractive indices (at wavelength 590 nm) of some better known crystals.

Many plastics are birefringent, because their molecules are 'frozen' in a stretched conformation when the plastic is molded or extruded. For example, ordinary cellophane is birefringent. Polarizers are routinely used to detect stress in plastics such as polystyrene and polycarbonate.

Cotton fiber is birefringent because of high levels of cellulosic material in the fiber's secondary cell wall.

Polarized light microscopy is commonly used in biological tissue, as many biological materials are birefringent. Collagen, found in cartilage, tendon, bone, corneas, and several other areas in the body, is birefringent and commonly studied with polarized light microscopy. Some proteins are also birefringent, exhibiting form birefringence.

Inevitable manufacturing imperfections in optical fiber leads to birefringence which is one cause of pulse broadening in fiber-optic communications. Such imperfections can be geometrical (lack of circular symmetry), due to stress applied to the optical fiber, and/or due to bending of the fiber. Birefringence is *intentionally* introduced (for instance, by making the cross-section elliptical) in order to produce polarization-maintaining optical fibers.

In addition to anisotropy in the electric polarizability (electric susceptibility), anisotropy in the magnetic polarizability (magnetic permeability) can also cause birefringence. However at optical frequencies, values of magnetic permeability for natural materials are not measurably different from μ_0 so this is not a source of optical birefringence in practice.

Uniaxial crystals, at 590 nm				
Material	**Crystal system**	n_o	n_e	Δn
barium borate BaB_2O_4	Trigonal	1.6776	1.5534	-0.1242
beryl $Be_3Al_2(SiO_3)_6$	Hexagonal	1.602	1.557	-0.045
calcite $CaCO_3$	Trigonal	1.658	1.486	-0.172
ice H_2O	Hexagonal	1.309	1.313	+0.004
lithium niobate $LiNbO_3$	Trigonal	2.272	2.187	-0.085
magnesium fluoride MgF_2	Tetragonal	1.380	1.385	+0.006

quartz SiO_2	Trigonal	1.544	1.553	+0.009
ruby Al_2O_3	Trigonal	1.770	1.762	-0.008
rutile TiO_2	Tetragonal	2.616	2.903	+0.287
sapphire Al_2O_3	Trigonal	1.768	1.760	-0.008
silicon carbide SiC	Hexagonal	2.647	2.693	+0.046
tourmaline (complex silicate)	Trigonal	1.669	1.638	-0.031
zircon, high $ZrSiO_4$	Tetragonal	1.960	2.015	+0.055
zircon, low $ZrSiO_4$	Tetragonal	1.920	1.967	+0.047

Biaxial crystals, at 590 nm				
Material	**Crystal system**	n_α	n_β	n_γ
borax $Na_2(B_4O_5)(OH)_4 \cdot 8(H_2O)$	Monoclinic	1.447	1.469	1.472
epsom salt $MgSO_4 \cdot 7(H_2O)$	Monoclinic	1.433	1.455	1.461
mica, biotite K(Mg,Fe) 3AlSi 3O 10(F,OH) 2	Monoclinic	1.595	1.640	1.640
mica, muscovite $KAl_2(AlSi_3O_{10})(F,OH)_2$	Monoclinic	1.563	1.596	1.601
olivine $(Mg, Fe)_2SiO_4$	Orthorhombic	1.640	1.660	1.680
perovskite $CaTiO_3$	Orthorhombic	2.300	2.340	2.380
topaz $Al_2SiO_4(F,OH)_2$	Orthorhombic	1.618	1.620	1.627
ulexite $NaCaB_5O_6(OH)_6 \cdot 5(H_2O)$	Triclinic	1.490	1.510	1.520

Measurement

Birefringence and other polarization based optical effects (such as optical rotation and linear or circular dichroism) can be measured by measuring the changes in the polarization of light passing through the material. These measurements are known as polarimetry. Polarized light microscopes, which contain two polarizers that are at 90 degrees to each other on either side of the sample, are used to visualize birefringence. The addition of quarter wave plates permit examination of circularly polarized light. Birefringence measurements have been made with phase-modulated systems for examining the transient flow behavior of fluids.

Birefringence of lipid bilayers can be measured using dual polarisation interferometry. This provides a measure of the degree of order within these fluid layers and how this order is disrupted when the layer interacts with other biomolecules.

Applications

Birefringence is used in many optical devices. Liquid crystal displays, the most common sort of flat panel display, cause their pixels to become lighter or darker through rotation of the polarization (circular birefringence) of linearly polarized light as viewed through a sheet polarizer at the screen's surface. Similarly, light modulators modulate the intensity of light through electrically induced birefringence of polarized light followed by a polarizer. The Lyot filter is a specialized narrowband spectral filter employing the wavelength dependence of birefringence. Wave plates are thin birefringent sheets widely used in certain optical equipment for modifying the polarization state of light passing through it.

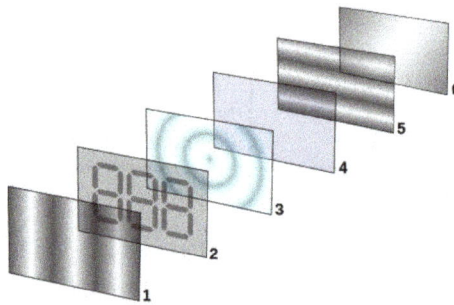

Reflective twisted nematic liquid crystal display. Light reflected by surface (6) (or coming from a backlight) is horizontally polarized (5) and passes through the liquid crystal modulator (3) sandwiched in between transparent layers (2,4) containing electrodes. Horizontally polarized light is blocked by the vertically oriented polarizer (1) except where its polarization has been rotated by the liquid crystal (3), appearing bright to the viewer

Birefringence also plays an important role in second harmonic generation and other nonlinear optical components, as the crystals used for this purpose are almost always birefringent. By adjusting the angle of incidence, the effective refractive index of the extraordinary ray can be tuned in order to achieve phase matching which is required for efficient operation of these devices.

Medicine

Birefringence is utilized in medical diagnostics. One powerful accessory used with optical microscopes is a pair of crossed polarizing filters. Light from the source is polarized in the X direction after passing through the first polarizer, but above the specimen is a polarizer (a so-called *analyzer*) oriented in the Y direction. Therefore, no light from the source will be accepted by the analyzer, and the field will appear dark. However areas of the sample possessing birefringence will generally couple some of the X polarized light into the Y polarization; these areas will then appear bright against the dark background. Modifications to this basic principle can differentiate between positive and negative birefringence.

For instance, needle aspiration of fluid from a gouty joint will reveal negatively birefringent monosodium urate crystals. Calcium pyrophosphate crystals, in contrast, show weak positive birefringence. Urate crystals appear yellow and calcium pyrophosphate crystals appear blue when their long axes are aligned parallel to that of a red compensator filter, or a crystal of known birefringence is added to the sample for comparison.

Birefringence can be observed in amyloid plaques such as are found in the brains of Alzheimer's patients when stained with a dye such as Congo Red. Modified proteins such as immunoglobulin light chains abnormally accumulate between cells, forming fibrils. Multiple folds of these fibers line up and take on a beta-pleated sheet conformation. Congo red dye intercalates between the folds and, when observed under polarized light, causes birefringence.

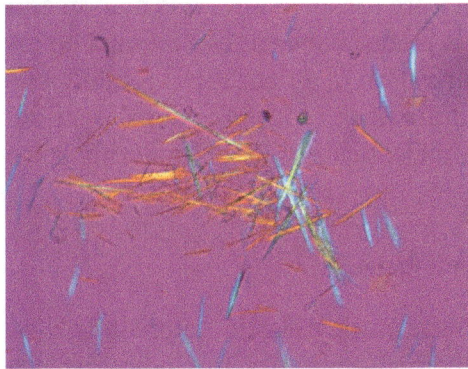

Urate crystals, with the crystals with their long axis seen as horizontal in this view being parallel to that of a red compensator filter. These appear as yellow, and are thereby of negative birefringence.

In ophthalmology, binocular retinal birefringence screening of the *Henle fibers* (photoreceptor axons that go radially outward from the fovea) provides a reliable detection of strabismus and possibly also of anisometropic amblyopia. Furthermore, scanning laser polarimetry utilises the birefringence of the optic nerve fibre layer to indirectly quantify its thickness, which is of use in the assessment and monitoring of glaucoma.

Birefringence characteristics in sperm heads allow for the selection of spermatozoa for intracytoplasmic sperm injection. Likewise, *zona imaging* uses birefringence on oocytes to select the ones with highest chances of successful pregnancy. Birefringence of particles biopsied from pulmonary nodules indicates silicosis.

Dermatologists use dermatascopes to view pigmented lesions and nevi. Dermatascopes use cross-polarized light, allowing the user to view chrystaline structures corresponding to dermal collagen in the skin. These structures may appear as shiny white lines or rosette shapes, and are only visible under polarized dermoscopy.

Stress Induced Birefringence

Color pattern of a plastic box with "frozen in" mechanical stress placed between two crossed polarizers.

Isotropic solids do not exhibit birefringence. However, when they are under mechanical stress, birefringence results. The stress can be applied externally or is "frozen in" after a birefringent plastic ware is cooled after it is manufactured using injection molding. When such a sample is placed between two crossed polarizers, colour patterns can be observed, because polarization of a light ray is rotated after passing through a birefingent material and the amount of rotation is dependent on wavelength. The experimental method called photoelasticity used for analyzing stress distribution in solids is based on the same principle.

Other Cases of Birefringence

Birefringence is observed in anisotropic elastic materials. In these materials, the two polarizations split according to their effective refractive indices which are also sensitive to stress.

The study of birefringence in shear waves traveling through the solid earth (the earth's liquid core does not support shear waves) is widely used in seismology.

Birefringence is widely used in mineralogy to identify rocks, minerals, and gemstones.

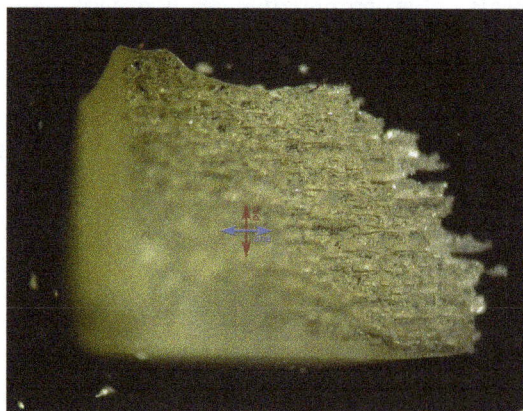

Birefringent rutile observed in different polarizations using a rotating polarizer (or analyzer)

Theory

In an isotropic medium (including free space) the so-called electric displacement (D) is just proportional to the electric field (E) according to $\mathbf{D} = \epsilon \mathbf{E}$ where the material's permittivity ε is just a scalar (and equal to the square of the index of refraction n times $\varepsilon 0$). However, in an anisotropic material exhibiting birefringence, the relationship between D and E must now be described using a tensor equation:

$$\mathbf{D} = \epsilon \mathbf{E}$$

where ε is now a 3x3 permittivity tensor. We assume linearity and no magnetic permeability in the medium: $\mu = \mu_0$. The electric field of a plane wave of angular frequency ω can be written in the general form:

$$\mathbf{E} = \mathbf{E_0} \exp\left[i(\mathbf{k} \cdot \mathbf{r} - \omega t)\right]$$

where r is the position vector, t is time, and E_0 is a vector describing the electric field at r=0, t=0. Then we shall find the possible wave vectors k. By combining Maxwell's equations for $\nabla \times E$ and $\nabla \times H$, we can eliminate $H = B / \mu_0$ to obtain:

$$-\nabla \times \nabla \times \mathbf{E} = \mu_0 \frac{\partial^2 \mathbf{D}}{\partial t^2} \quad \text{3a}$$

With no free charges, Maxwell's equation for the divergence of D vanishes:

$$\nabla \cdot \mathbf{D} = 0 \ (3b)$$

We can apply the vector identity:

$$\nabla \times \left(\nabla \times \mathbf{A}\right) = \nabla(\nabla \cdot \mathbf{A}) - \nabla^2 \mathbf{A}$$

to the left hand side of eqn. 3a, and use the spatial dependence in which each differentiation in x (for instance) results in multiplication by ik_x to find:

$$-\nabla \times \nabla \times \mathbf{E} = (\mathbf{k} \cdot \mathbf{E})\mathbf{k} - (\mathbf{k} \cdot \mathbf{k})\mathbf{E}.$$

The right hand side of eqn. 3a can be expressed in terms of E through application of the permitivity tensor ε and noting that differentiation in time results in multiplication by $-i\omega$, eqn. 3a then becomes:

$$(-\mathbf{k} \cdot \mathbf{k})\mathbf{E} + (\mathbf{k} \cdot \mathbf{E})\mathbf{k} = -\mu_0 \omega^2 (\epsilon \mathbf{E}) \qquad (4a)$$

Applying the differentiation rule to eqn. 3b we find:

$$\mathbf{k} \cdot \mathbf{D} = 0 \qquad (4b)$$

Eqn. 4b indicates that $\mathbf{D}$ is orthogonal to the direction of the wavevector k, even though that is no longer generally true for **E** as would be the case in an iso tropic medium. Eqn. 4b will not be needed for the further steps in the following derivation.

Finding the allowed values of k for a given ω is easiest done by using Cartesian coordinates with the x, y and z axes chosen in the directions of the symmetry axes of the crystal (or simply choosing z in the direction of the optic axis of a uniaxial crystal), resulting in a diagonal matrix for the permitivity tensor ε:

$$\epsilon = \epsilon_0 \begin{bmatrix} n_x^2 & 0 & 0 \\ 0 & n_y^2 & 0 \\ 0 & 0 & n_z^2 \end{bmatrix}$$

(4c)

where the diagonal values are squares of the refractive indices for polarizations along the three principal axes x, y and z. With ε in this form, and substituting in the speed of light c using $c = 1/\sqrt{\mu_0 \dot{\theta_0}}$, eqn. 4a becomes

$$\left(-k_x^2 - k_y^2 - k_z^2\right)E_x + k_x^2 E_x + k_x k_y E_y + k_x k_z E_z = -\frac{\omega^2 n_x^2}{c^2} E_x$$

(5a)

where E_x, E_y, E_z are the components of **E** (at any given position in space and time) and k_x, k_y and k_z are the components of k. Rearranging, we can write (and similarly for the y and z components of eqn. 4a):

$$\left(-k_y^2 - k_z^2 + \frac{\omega^2 n_x^2}{c^2}\right)E_x + k_x k_y E_y + k_x k_z E_z = 0$$

(5b)

$$k_x k_y E_x + \left(-k_x^2 - k_z^2 + \frac{\omega^2 n_y^2}{c^2}\right)E_y + k_y k_z E_z = 0$$

(5c)

$$k_x k_z E_x + k_y k_z E_y + \left(-k_x^2 - k_y^2 + \frac{\omega^2 n_z^2}{c^2}\right)E_z = 0$$

(5d)

This is a set of linear equations in E_x, E_y, E_z, so it can have a non-trivial solution (that is, one other than $E=0$) as long as the following determinant is zero:

$$\begin{vmatrix} \left(-k_y^2 - k_z^2 + \dfrac{\omega^2 n_x^2}{c^2}\right) & k_x k_y & k_x k_z \\[4mm] k_x k_y & \left(-k_x^2 - k_z^2 + \dfrac{\omega^2 n_y^2}{c^2}\right) & k_y k_z \\[4mm] k_x k_z & k_y k_z & \left(-k_x^2 - k_y^2 + \dfrac{\omega^2 n_z^2}{c^2}\right) \end{vmatrix} = 0 \tag{6}$$

Evaluating the determinant of eqn (6), and rearranging the terms, we obtain

$$\frac{\omega^4}{c^4} - \frac{\omega^2}{c^2}\left(\frac{k_x^2 + k_y^2}{n_z^2} + \frac{k_x^2 + k_z^2}{n_y^2} + \frac{k_y^2 + k_z^2}{n_x^2}\right) +$$

$$\left(\frac{k_x^2}{n_y^2 n_z^2} + \frac{k_y^2}{n_x^2 n_z^2} + \frac{k_z^2}{n_x^2 n_y^2}\right)(k_x^2 + k_y^2 + k_z^2) = 0$$

In the case of a uniaxial material, choosing the optic axis to be in the z direction so that $n_x = n_y = n_o$ and $n_z = n_e$, this expression can be factored into

$$\left(\frac{k_x^2}{n_o^2} + \frac{k_y^2}{n_o^2} + \frac{k_z^2}{n_o^2} - \frac{\omega^2}{c^2}\right)\left(\frac{k_x^2}{n_e^2} + \frac{k_y^2}{n_e^2} + \frac{k_z^2}{n_o^2} - \frac{\omega^2}{c^2}\right) = 0 \tag{8}$$

Setting either of the factors in equation (8) to zero will define an ellipsoidal surface in the space of wave vectors k that are allowed for a given ω. The first factor being zero defines a *sphere*; this is the solution for "ordinary rays", in which the effective refractive index is exactly n_o regardless of the direction of k. The second defines a spheroid symmetric about the z axis. This solution corresponds to the "extraordinary rays" in which the effective refractive index is in between n_o and n_e, depending on the direction of k. Therefore, for any arbitrary direction of propagation (other than in the direction of the optic axis), two distinct wavevectors k are allowed corresponding to the polarizations of the ordinary and extraordinary rays.

For a biaxial material a similar but more complicated condition on the two waves can be described; the allowed k vectors in a specified direction now lie on one of two ellipsoids. By inspection one can see that eqn. 6 is generally satisfied for two positive values of ω. Or for a specified optical frequency ω and direction normal to the wavefronts k/|k|, it is satisfied for two wavenumbers (propagation constants) |k| (and thus effective refractive indices) corresponding to the propagation of two linear polarizations in that direction.

When those two propagation constants are equal then the effective refractive index is independent of polarization, and there is consequently *no* birefringence encountered by a wave traveling in that particular direction. For a uniaxial crystal, this is the *optic axis*, the z direction according to the above construction. But when all three refractive indices (or permittivities), n_x, n_y and n_z are distinct, it can be shown that there are exactly two such directions (where the two ellipsoids intersect); these directions are not at all obvious and do *not* lie along any of the three principal axes (*x*, *y*, and *z* according to the above convention). Historically that accounts for the use of the term "biaxial" for such crystals, as the existence of exactly two such special directions (considered "axes") was discovered well before polarization and birefringence were understood physically. However these two special directions are generally not of particular interest; biaxial crystals are rather specified by their three refractive indices corresponding to the three axes of symmetry.

A general state of polarization launched into the medium can always be decomposed into two waves, one in each of those two polarizations, which will then propagate with different wavenumbers |k|. Applying the different phase of propagation to those two waves over a specified propagation distance will result in a generally *different* net polarization state at that point; this is the principle of the waveplate for instance. However, when you have a wave launched into a birefringent material at non-normal incidence, the problem is yet more complicated since the two polarization components will now not only have distinct wavenumbers but the k vectors will not even be in exactly the same direction. In this case the two k vectors are rather solutions of eqn. 6 constrained by the boundary condition which requires that the components of the two transmitted waves' k vectors, and the k vector of the incident wave, as *projected onto the surface of the interface* must all be identical.

References

- Lekner, John (1987). Theory of Reflection, of Electromagnetic and Particle Waves. Springer. ISBN 9789024734184.

- Bohren, Craig F.; Donald R. Huffman (1983). Absorption and Scattering of Light by Small Particles. Wiley. ISBN 0-471-29340-7.

- Colton, David; Rainer Kress (1998). Inverse Acoustic and Electromagnetic Scattering Theory. Springer. ISBN 3-540-62838-X.

- Seinfeld, John H.; Pandis, Spyros N. (2006). Atmospheric Chemistry and Physics - From Air Pollution to Climate Change (2nd Ed.). John Wiley and Sons, Inc. ISBN 0-471-82857-2

- Stover, John C. (1995). Optical Scattering: Measurement and Analysis. SPIE Optical Engineering Press. ISBN 0-8194-1934-6.

- Greene, Brian (1999). The Elegant Universe: Superstrings, Hidden Dimensions, and the Quest for the Ultimate Theory. New York: W.W. Norton. pp. 97–109. ISBN 0-393-04688-5.

Design and Analytical Aspects of Optical Instrument

This chapter will provide an integrated understanding of the design and analytical aspects of optical instrument. This chapter educates the reader on concepts as lens, prism, microscope, telescope, photometer, polarimeter, lensmeter, waveplate etc. The chapter strategically encompasses and incorporates the major components and key concepts of optical engineering, providing a complete understanding.

Lens (Optics)

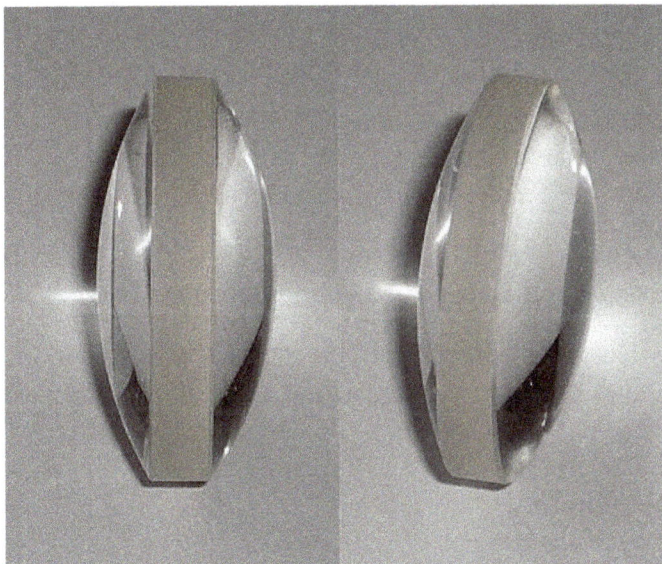

A biconvex lens

A lens is a transmissive optical device that focuses or disperses a light beam by means of refraction. A simple lens consists of a single piece of transparent material, while a compound lens consists of several simple lenses (*elements*), usually arranged along a common axis. Lenses are made from materials such as glass or plastic, and are ground and polished or moulded to a desired shape. A lens can focus light to form an image, unlike a prism, which refracts light without focusing. Devices that similarly focus or disperse radiation other than visible light are also called lenses, such as microwave lenses, electron lenses or acoustic lenses.

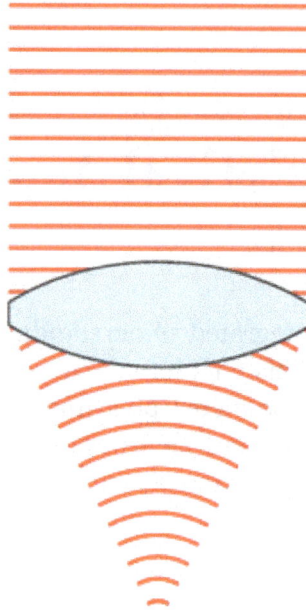

Lenses can be used to focus light

History

The word *lens* comes from the Latin name of the lentil, because a double-convex lens is lentil-shaped. The genus of the lentil plant is *Lens*, and the most commonly eaten species is *Lens culinaris*. The lentil plant also gives its name to a geometric figure.

The variant spelling *lense* is sometimes seen. While it is listed as an alternative spelling in some dictionaries, most mainstream dictionaries do not list it as acceptable.

The oldest lens artifact is the Nimrud lens, dating back 2700 years (7th century B.C.) to ancient Assyria. David Brewster proposed that it may have been used as a magnifying glass, or as a burning-glass to start fires by concentrating sunlight. Another early reference to magnification dates back to ancient Egyptian hieroglyphs in the 8th century BC, which depict "simple glass meniscal lenses".

The earliest written records of lenses date to Ancient Greece, with Aristophanes' play *The Clouds* (424 BC) mentioning a burning-glass (a biconvex lens used to focus the sun's rays to produce fire). Some scholars argue that the archeological evidence indicates that there was widespread use of lenses in antiquity, spanning several millennia. Such lenses were used by artisans for fine work, and for authenticating seal impressions. The writings of Pliny the Elder (23–79) show that burning-glasses were known to the Roman Empire, and mentions what is arguably the earliest written reference to a corrective lens: Nero was said to watch the gladiatorial games using an emerald (presumably concave to correct for nearsightedness, though the reference is vague). Both Pliny and Seneca the Younger (3 BC–65) described the magnifying effect of a glass globe filled with water.

Excavations at the Viking harbour town of Fröjel, Gotland, Sweden discovered in 1999 the rock crystal Visby lenses, produced by turning on pole lathes at Fröjel in the 11th to 12th century, with an imaging quality comparable to that of 1950s aspheric lenses. The Viking lenses were capable of concentrating enough sunlight to ignite fires.

Between the 11th and 13th century "reading stones" were invented. Often used by monks to assist in illuminating manuscripts, these were primitive plano-convex lenses initially made by cutting a glass sphere in half. As the stones were experimented with, it was slowly understood that shallower lenses magnified more effectively.

Lenses came into widespread use in Europe with the invention of spectacles, probably in Italy in the 1280s. This was the start of the optical industry of grinding and polishing lenses for spectacles, first in Venice and Florence in the thirteenth century, and later in the spectacle-making centres in both the Netherlands and Germany. Spectacle makers created improved types of lenses for the correction of vision based more on empirical knowledge gained from observing the effects of the lenses (probably without the knowledge of the rudimentary optical theory of the day). The practical development and experimentation with lenses led to the invention of the compound optical microscope around 1595, and the refracting telescope in 1608, both of which appeared in the spectacle-making centres in the Netherlands.

With the invention of the telescope and microscope there was a great deal of experimentation with lens shapes in the 17th and early 18th centuries trying to correct chromatic errors seen in lenses. Opticians tried to construct lenses of varying forms of curvature, wrongly assuming errors arose from defects in the spherical figure of their surfaces. Optical theory on refraction and experimentation was showing no single-element lens could bring all colours to a focus. This led to the invention of the compound achromatic lens by Chester Moore Hall in England in 1733, an invention also claimed by fellow Englishman John Dollond in a 1758 patent.

Construction of Simple Lenses

Most lenses are *spherical lenses*: their two surfaces are parts of the surfaces of spheres. Each surface can be *convex* (bulging outwards from the lens), *concave* (depressed into the lens), or *planar* (flat). The line joining the centres of the spheres making up the lens surfaces is called the *axis* of the lens. Typically the lens axis passes through the physical centre of the lens, because of the way they are manufactured. Lenses may be cut or ground after manufacturing to give them a different shape or size. The lens axis may then not pass through the physical centre of the lens.

Toric or sphero-cylindrical lenses have surfaces with two different radii of curvature in two orthogonal planes. They have a different focal power in different meridians. This forms an astigmatic lens. An example is eyeglass lenses that are used to correct astigmatism in someone's eye.

More complex are aspheric lenses. These are lenses where one or both surfaces have a shape that is neither spherical nor cylindrical. The more complicated shapes allow such lenses to form images with less aberration than standard simple lenses, but they are more difficult and expensive to produce.

Types of Simple Lenses

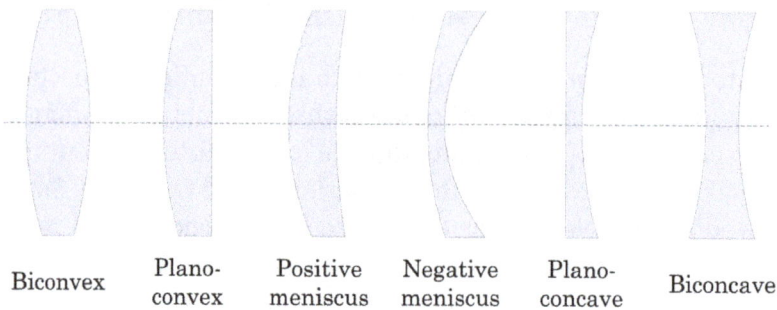

| Biconvex | Plano-convex | Positive meniscus | Negative meniscus | Plano-concave | Biconcave |

Lenses are classified by the curvature of the two optical surfaces. A lens is *biconvex* (or *double convex*, or just *convex*) if both surfaces are convex. If both surfaces have the same radius of curvature, the lens is *equiconvex*. A lens with two concave surfaces is *biconcave* (or just *concave*). If one of the surfaces is flat, the lens is *plano-convex* or *plano-concave* depending on the curvature of the other surface. A lens with one convex and one concave side is *convex-concave* or *meniscus*. It is this type of lens that is most commonly used in corrective lenses.

If the lens is biconvex or plano-convex, a collimated beam of light passing through the lens converges to a spot (a *focus*) behind the lens. In this case, the lens is called a *positive* or *converging* lens. The distance from the lens to the spot is the focal length of the lens, which is commonly abbreviated f in diagrams and equations.

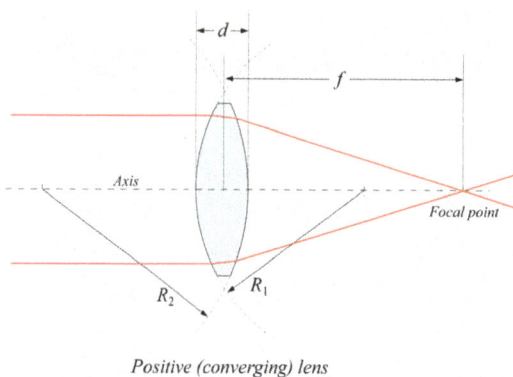

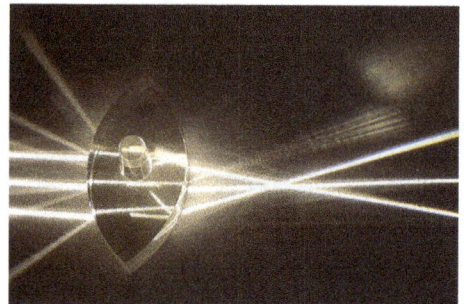

Positive (converging) lens

If the lens is biconcave or plano-concave, a collimated beam of light passing through the lens is diverged (spread); the lens is thus called a *negative* or *diverging* lens. The beam, after passing through the lens, appears to emanate from a particular point on the axis in front of the lens. The distance from this point to the lens is also known as the focal length, though it is negative with respect to the focal length of a converging lens.

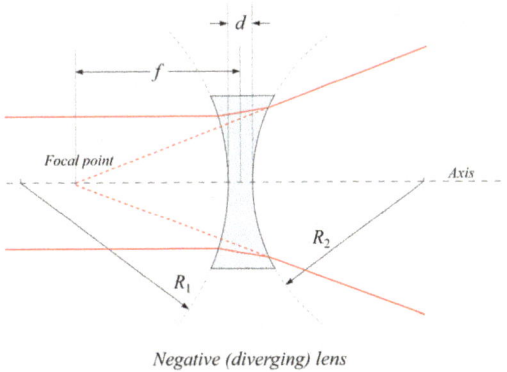

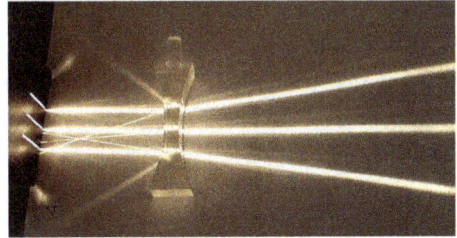

Negative (diverging) lens

Convex-concave (meniscus) lenses can be either positive or negative, depending on the relative curvatures of the two surfaces. A *negative meniscus* lens has a steeper concave surface and is thinner at the centre than at the periphery. Conversely, a *positive meniscus* lens has a steeper convex surface and is thicker at the centre than at the periphery. An ideal thin lens with two surfaces of equal curvature would have zero optical power, meaning that it would neither converge nor diverge light. All real lenses have nonzero thickness, however, which makes a real lens with identical curved surfaces slightly positive. To obtain exactly zero optical power, a meniscus lens must have slightly unequal curvatures to account for the effect of the lens' thickness.

Lensmaker's Equation

The focal length of a lens *in air* can be calculated from the lensmaker's equation:

$$\frac{1}{f} = (n-1)\left[\frac{1}{R_1} - \frac{1}{R_2} + \frac{(n-1)d}{nR_1R_2}\right],$$

where

f is the focal length of the lens,

n is the refractive index of the lens material,

R_1 is the radius of curvature of the lens surface closest to the light source,

R_2 is the radius of curvature of the lens surface farthest from the light source, and

d is the thickness of the lens (the distance along the lens axis between the two surface vertices).

The focal length f is positive for converging lenses, and negative for diverging lenses. The reciprocal of the focal length, $1/f$, is the optical power of the lens. If the focal length is in metres, this gives the optical power in dioptres (inverse metres).

Lenses have the same focal length when light travels from the back to the front as when light goes from the front to the back. Other properties of the lens, such as the aberrations are not the same in both directions.

Sign Convention for Radii of Curvature R_1 and R_2

The signs of the lens' radii of curvature indicate whether the corresponding surfaces are convex or concave. The sign convention used to represent this varies, but in this article a *positive R* indicates a surface's center of curvature is further along in the direction of the ray travel (right, in the accompanying diagrams), while *negative R* means that rays reaching the surface have already passed the center of curvature. Consequently, for external lens surfaces as diagrammed above, $R1 > 0$ and $R2 < 0$ indicate *convex* surfaces (used to converge light in a positive lens), while $R1 < 0$ and $R2 > 0$ indicate *concave* surfaces. The reciprocal of the radius of curvature is called the curvature. A flat surface has zero curvature, and its radius of curvature is infinity.

Thin Lens Approximation

If d is small compared to R_1 and R_2, then the *thin lens* approximation can be made. For a lens in air, f is then given by

$$\frac{1}{f} \approx (n-1)\left[\frac{1}{R_1} - \frac{1}{R_2}\right].$$

Imaging Properties

As mentioned above, a positive or converging lens in air focuses a collimated beam travelling along the lens axis to a spot (known as the focal point) at a distance f from the lens. Conversely, a point source of light placed at the focal point is converted into a collimated beam by the lens. These two cases are examples of image formation in lenses. In the former case, an object at an infinite distance (as represented by a collimated beam of waves) is focused to an image at the focal point of the lens. In the latter, an object at the focal length distance from the lens is imaged at infinity. The plane perpendicular to the lens axis situated at a distance f from the lens is called the *focal plane*.

If the distances from the object to the lens and from the lens to the image are S_1 and S_2

respectively, for a lens of negligible thickness, in air, the distances are related by the thin lens formula:

$$\frac{1}{S_1} + \frac{1}{S_2} = \frac{1}{f}.$$

This can also be put into the "Newtonian" form:

$$x_1 x_2 = f^2,$$

where $x_1 = S_1 - f$ and $x_2 = S_2 - f$.

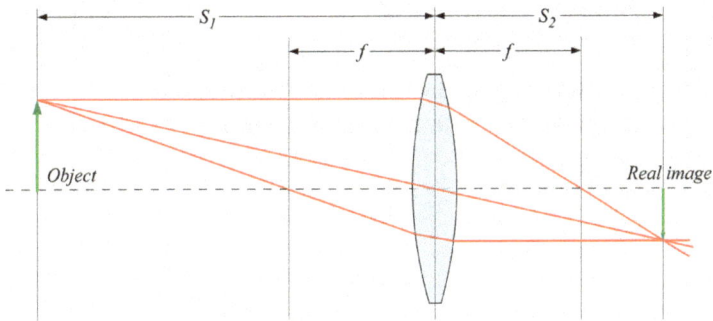

A camera lens forms a real image of a distant object.

Therefore, if an object is placed at a distance $S_1 > f$ from a positive lens of focal length f, we will find an image distance S_2 according to this formula. If a screen is placed at a distance S_2 on the opposite side of the lens, an image is formed on it. This sort of image, which can be projected onto a screen or image sensor, is known as a *real image*.

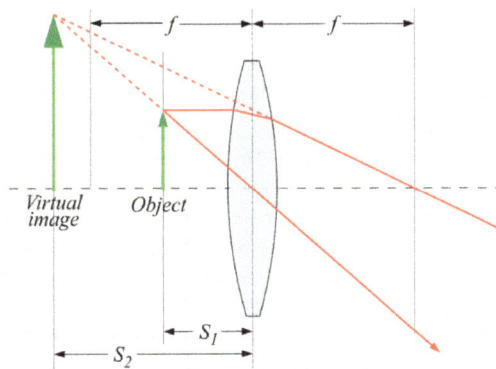

Virtual image formation using a positive lens as a magnifying glass.

This is the principle of the camera, and of the human eye. The focusing adjustment of a camera adjusts S_2, as using an image distance different from that required by this formula produces a defocused (fuzzy) image for an object at a distance of S_1 from the camera. Put another way, modifying S_2 causes objects at a different S_1 to come into perfect focus.

In some cases S_2 is negative, indicating that the image is formed on the opposite side of the lens from where those rays are being considered. Since the diverging light rays emanating from the lens never come into focus, and those rays are not physically present at the point where they *appear* to form an image, this is called a virtual image. Unlike real images, a virtual image cannot be projected on a screen, but appears to an observer looking through the lens as if it were a real object at the location of that virtual image. Likewise, it appears to a subsequent lens as if it were an object at that location, so that second lens could again focus that light into a real image, S_1 then being measured from the virtual image location behind the first lens to the second lens. This is exactly what the eye does when looking through a magnifying glass. The magnifying glass creates a (magnified) virtual image behind the magnifying glass, but those rays are then re-imaged by the lens of the eye to create a *real image* on the retina.

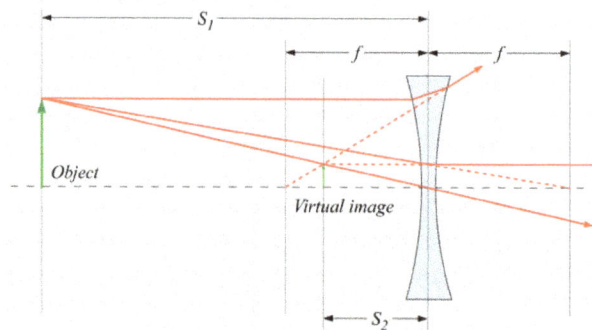

A *negative* lens produces a demagnified virtual image.

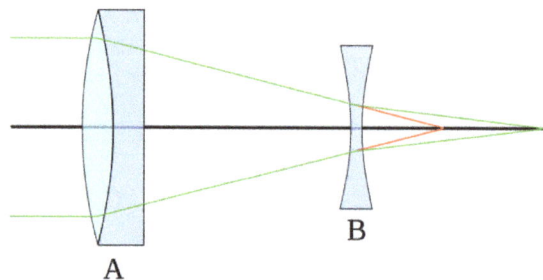

A Barlow lens (B) reimages a virtual object (focus of red ray path) into a magnified real image (green rays at focus)

Using a positive lens of focal length f, a virtual image results when $S_1 < f$, the lens thus being used a magnifying glass (rather than if $S_1 \gg f$ as for a camera). Using a negative lens ($f < 0$) with a *real object* ($S_1 > 0$) can only produce a virtual image ($S_2 < 0$), according to the above formula. It is also possible for the object distance S_1 to be negative, in which case the lens sees a so-called *virtual object*. This happens when the lens is inserted into a converging beam (being focused by a previous lens) *before* the location of its real image. In that case even a negative lens can project a real image, as is done by a Barlow lens.

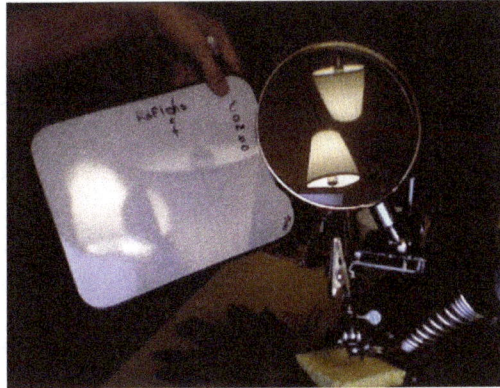

Real image of a lamp is projected onto a screen (inverted). Reflections of the lamp from both surfaces of the biconvex lens are visible.

A convex lens ($f << S_1$) forming a real, inverted image rather than the upright, virtual image as seen in a magnifying glass

For a thin lens, the distances S_1 and S_2 are measured from the object and image to the position of the lens, as described above. When the thickness of the lens is not much smaller than S_1 and S_2 or there are multiple lens elements (a compound lens), one must instead measure from the object and image to the principal planes of the lens. If distances S_1 or S_2 pass through a medium other than air or vacuum a more complicated analysis is required.

Magnification

The linear *magnification* of an imaging system using a single lens is given by

$$M = -\frac{S_2}{S_1} = \frac{f}{f - S_1},$$

where M is the magnification factor defined as the ratio of the size of an image compared to the size of the object. The sign convention here dictates that if M is negative, as it is for real images, the image is upside-down with respect to the object. For virtual images M is positive, so the image is upright.

Linear magnification M is not always the most useful measure of magnifying power. For instance, when characterizing a visual telescope or binoculars that produce only a virtual image, one would be more concerned with the angular magnification—which expresses how much larger a distant object appears through the telescope compared to the naked eye. In the case of a camera one would quote the plate scale, which compares the apparent (angular) size of a distant object to the size of the real image produced at the focus. The plate scale is the reciprocal of the focal length of the camera lens; lenses are categorized as long-focus lenses or wide-angle lenses according to their focal lengths.

Using an inappropriate measurement of magnification can be formally correct but yield a meaningless number. For instance, using a magnifying glass of 5 cm focal length, held 20 cm from the eye and 5 cm from the object, produces a virtual image at infinity of infinite linear size: $M = \infty$. But the *angular magnification* is 5, meaning that the object appears 5 times larger to the eye than without the lens. When taking a picture of the moon using a camera with a 50 mm lens, one is not concerned with the linear magnification $M \approx -50$ mm / 380000 km $= -1.3 \times 10{-}10$. Rather, the plate scale of the camera is about 1°/mm, from which one can conclude that the 0.5 mm image on the film corresponds to an angular size of the moon seen from earth of about 0.5°.

In the extreme case where an object is an infinite distance away, $S_1 = \infty$, $S_2 = f$ and $M = -f/\infty = 0$, indicating that the object would be imaged to a single point in the focal plane. In fact, the diameter of the projected spot is not actually zero, since diffraction places a lower limit on the size of the point spread function. This is called the diffraction limit.

Aberrations

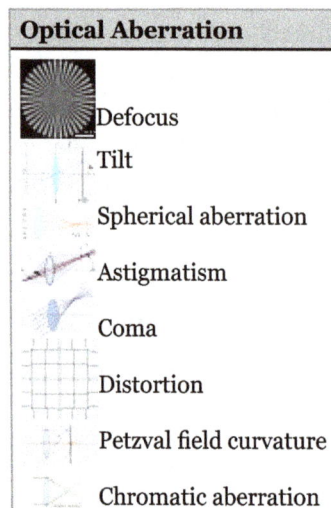

Optical Aberration

Defocus

Tilt

Spherical aberration

Astigmatism

Coma

Distortion

Petzval field curvature

Chromatic aberration

Lenses do not form perfect images, and a lens always introduces some degree of distortion or *aberration* that makes the image an imperfect replica of the object. Careful design of the lens system for a particular application minimizes the aberration. Several types of aberration affect image quality, including spherical aberration, coma, and chromatic aberration.

Spherical Aberration

Spherical aberration occurs because spherical surfaces are not the ideal shape for a lens, but are by far the simplest shape to which glass can be ground and polished, and so are often used. Spherical aberration causes beams parallel to, but distant from, the lens axis to be focused in a slightly different place than beams close to the axis. This manifests itself as a blurring of the image. Lenses in which closer-to-ideal, non-spherical surfaces are used are called *aspheric* lenses. These were formerly complex to make and often extremely expensive, but advances in technology have greatly reduced the manufacturing cost for such lenses. Spherical aberration can be minimised by carefully choosing the surface curvatures for a particular application. For instance, a plano-convex lens, which is used to focus a collimated beam, produces a sharper focal spot when used with the convex side towards the beam source.

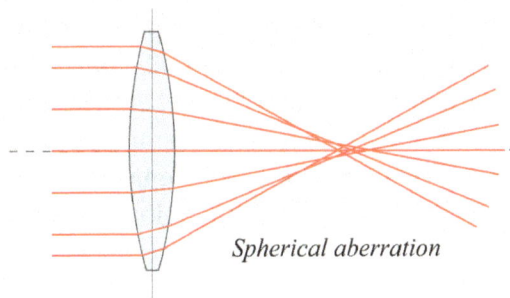

Spherical aberration

Coma

Coma, or *comatic aberration*, derives its name from the comet-like appearance of the aberrated image. Coma occurs when an object off the optical axis of the lens is imaged, where rays pass through the lens at an angle to the axis θ. Rays that pass through the centre of a lens of focal length f are focused at a point with distance f tan θ from the axis. Rays passing through the outer margins of the lens are focused at different points, either further from the axis (positive coma) or closer to the axis (negative coma). In general, a bundle of parallel rays passing through the lens at a fixed distance from the centre of the lens are focused to a ring-shaped image in the focal plane, known as a *comatic circle*. The sum of all these circles results in a V-shaped or comet-like flare. As with spherical aberration, coma can be minimised (and in some cases eliminated) by choosing the curvature of the two lens surfaces to match the application. Lenses in which both spherical aberration and coma are minimised are called *bestform* lenses.

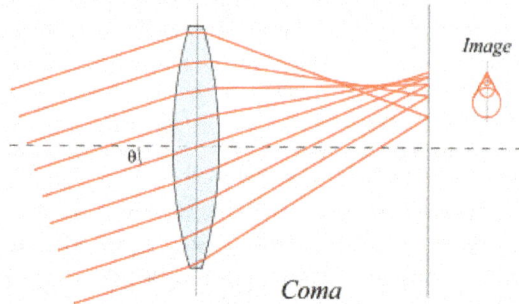

Coma

Chromatic Aberration

Chromatic aberration is caused by the dispersion of the lens material—the variation of its refractive index, n, with the wavelength of light. Since, from the formulae above, f is dependent upon n, it follows that light of different wavelengths is focused to different positions. Chromatic aberration of a lens is seen as fringes of colour around the image. It can be minimised by using an achromatic doublet (or *achromat*) in which two materials with differing dispersion are bonded together to form a single lens. This reduces the amount of chromatic aberration over a certain range of wavelengths, though it does not produce perfect correction. The use of achromats was an important step in the development of the optical microscope. An apochromat is a lens or lens system with even better chromatic aberration correction, combined with improved spherical aberration correction. Apochromats are much more expensive than achromats.

Different lens materials may also be used to minimise chromatic aberration, such as specialised coatings or lenses made from the crystal fluorite. This naturally occurring substance has the highest known Abbe number, indicating that the material has low dispersion.

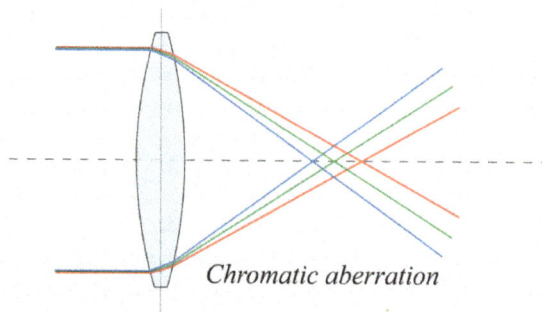

Chromatic aberration

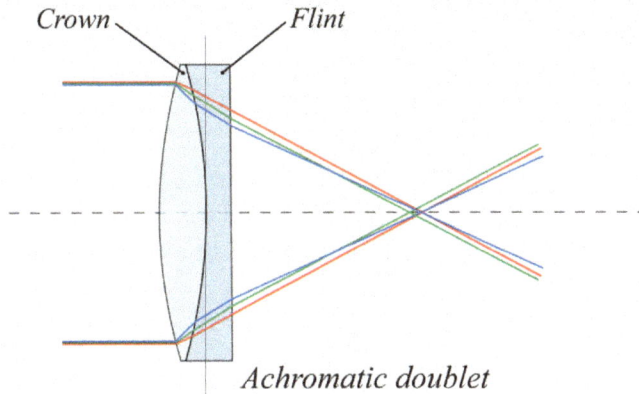

Achromatic doublet

Other Types of Aberration

Other kinds of aberration include *field curvature, barrel* and *pincushion distortion,* and *astigmatism.*

Aperture Diffraction

Even if a lens is designed to minimize or eliminate the aberrations described above, the image quality is still limited by the diffraction of light passing through the lens' finite aperture. A diffraction-limited lens is one in which aberrations have been reduced to the point where the image quality is primarily limited by diffraction under the design conditions.

Compound Lenses

Simple lenses are subject to the optical aberrations discussed above. In many cases these aberrations can be compensated for to a great extent by using a combination of simple lenses with complementary aberrations. A *compound lens* is a collection of simple lenses of different shapes and made of materials of different refractive indices, arranged one after the other with a common axis.

The simplest case is where lenses are placed in contact: if the lenses of focal lengths f_1 and f_2 are "thin", the combined focal length f of the lenses is given by

$$\frac{1}{f} = \frac{1}{f_1} + \frac{1}{f_2}.$$

Since $1/f$ is the power of a lens, it can be seen that the powers of thin lenses in contact are additive.

If two thin lenses are separated in air by some distance d, the focal length for the combined system is given by

$$\frac{1}{f} = \frac{1}{f_1} + \frac{1}{f_2} - \frac{d}{f_1 f_2}.$$

The distance from the front focal point of the combined lenses to the first lens is called the *front focal length* (FFL):

$$FFL = \frac{f_1(f_2 - d)}{(f_1 + f_2) - d}.$$

Similarly, the distance from the second lens to the rear focal point of the combined system is the *back focal length* (BFL):

$$BFL = \frac{f_2(d - f_1)}{d - (f_1 + f_2)}.$$

As d tends to zero, the focal lengths tend to the value of f given for thin lenses in contact.

If the separation distance is equal to the sum of the focal lengths ($d = f_1 + f_2$), the FFL and BFL are infinite. This corresponds to a pair of lenses that transform a parallel (collimated) beam into another collimated beam. This type of system is called an *afocal system*, since it produces no net convergence or divergence of the beam. Two lenses at this separation form the simplest type of optical telescope. Although the system does not alter the divergence of a collimated beam, it does alter the width of the beam. The magnification of such a telescope is given by

$$M = -\frac{f_2}{f_1},$$

which is the ratio of the output beam width to the input beam width. Note the sign convention: a telescope with two convex lenses ($f_1 > 0, f_2 > 0$) produces a negative magnification, indicating an inverted image. A convex plus a concave lens ($f_1 > 0 > f_2$) produces a positive magnification and the image is upright.

Other Types

Cylindrical lenses have curvature in only one direction. They are used to focus light into a line, or to convert the elliptical light from a laser diode into a round beam.

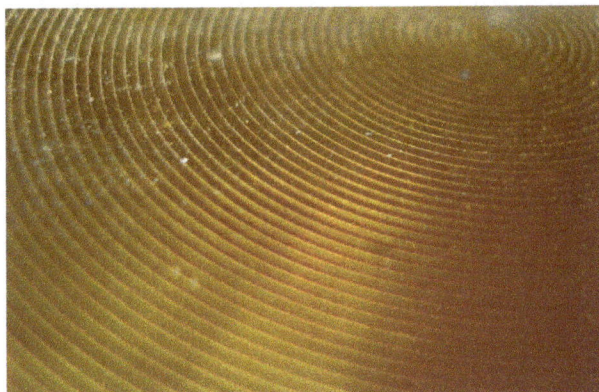

Close-up view of a flat Fresnel lens.

A Fresnel lens has its optical surface broken up into narrow rings, allowing the lens to be much thinner and lighter than conventional lenses. Durable Fresnel lenses can be molded from plastic and are inexpensive.

Lenticular lenses are arrays of microlenses that are used in lenticular printing to make images that have an illusion of depth or that change when viewed from different angles.

A gradient index lens has flat optical surfaces, but has a radial or axial variation in index of refraction that causes light passing through the lens to be focused.

An axicon has a conical optical surface. It images a point source into a line *along* the optic axis, or transforms a laser beam into a ring.

Diffractive optical elements can function as lenses.

Superlenses are made from negative index metamaterials and claim to produce images at spatial resolutions exceeding the diffraction limit. The first superlenses were made in 2004 using such a metamaterial for microwaves. Improved versions have been made by other researchers.As of 2014 the superlens has not yet been demonstrated at visible or near-infrared wavelengths.

A prototype flat ultrathin lens, with no curvature has been developed.

Uses

A single convex lens mounted in a frame with a handle or stand is a magnifying glass.

Lenses are used as prosthetics for the correction of visual impairments such as myopia, hyperopia, presbyopia, and astigmatism. Most lenses used for other purposes have strict axial symmetry; eyeglass lenses are only approximately symmetric. They are usually shaped to fit in a roughly oval, not circular, frame; the optical centres are placed over the eyeballs; their curvature may not be axially symmetric to correct for astigmatism. Sunglasses' lenses are designed to attenuate light; sunglass lenses that also correct visual impairments can be custom made.

Other uses are in imaging systems such as monoculars, binoculars, telescopes, microscopes, cameras and projectors. Some of these instruments produce a virtual image when applied to the human eye; others produce a real image that can be captured on photographic film or an optical sensor, or can be viewed on a screen. In these devices lenses are sometimes paired up with curved mirrors to make a catadioptric system where the lens's spherical aberration corrects the opposite aberration in the mirror (such as Schmidt and meniscus correctors).

Convex lenses produce an image of an object at infinity at their focus; if the sun is imaged, much of the visible and infrared light incident on the lens is concentrated into the small image. A large lens creates enough intensity to burn a flammable object at the focal point. Since ignition can be achieved even with a poorly made lens, lenses have been used as burning-glasses for at least 2400 years. A modern application is the use of relatively large lenses to concentrate solar energy on relatively small photovoltaic cells, harvesting more energy without the need to use larger and more expensive cells.

Radio astronomy and radar systems often use dielectric lenses, commonly called a lens antenna to refract electromagnetic radiation into a collector antenna.

Lenses can become scratched and abraded. Abrasion-resistant coatings are available to help control this.

Prism

A plastic prism

In optics, a prism is a transparent optical element with flat, polished surfaces that refract light. At least two of the flat surfaces must have an angle between them. The exact angles between the surfaces depend on the application. The traditional geometrical shape is that of a triangular prism with a triangular base and rectangular sides, and in colloquial use "prism" usually refers to this type. Some types of optical prism are not in fact in the shape of geometric prisms. Prisms can be made from any material that is transparent to the wavelengths for which they are designed. Typical materials include glass, plastic and fluorite.

A dispersive prism can be used to break light up into its constituent spectral colors (the colors of the rainbow). Furthermore, prisms can be used to reflect light, or to split light into components with different polarizations.

How Prisms Work

Light changes speed as it moves from one medium to another (for example, from air into the glass of the prism). This speed change causes the light to be refracted and to enter the new medium at a different angle (Huygens principle). The degree of bending of the light's path depends on the angle that the incident beam of light makes with the surface, and on the ratio between the refractive indices of the two media (Snell's law). The refractive index of many materials (such as glass) varies with the wavelength or color of the light used, a phenomenon known as *dispersion*. This causes light of different colors to be refracted differently and to leave the prism at different angles, creating an effect similar to a rainbow. This can be used to separate a beam of white light into its constituent spectrum of colors. Prisms will generally disperse light over a much larger frequency bandwidth than diffraction gratings, making them useful for broad-spectrum spectroscopy. Furthermore, prisms do not suffer from complications arising from overlapping spectral orders, which all gratings have.

Prisms are sometimes used for the internal reflection at the surfaces rather than for dispersion. If light inside the prism hits one of the surfaces at a sufficiently steep angle, total internal reflection occurs and *all* of the light is reflected. This makes a prism a useful substitute for a mirror in some situations.

Deviation Angle and Dispersion

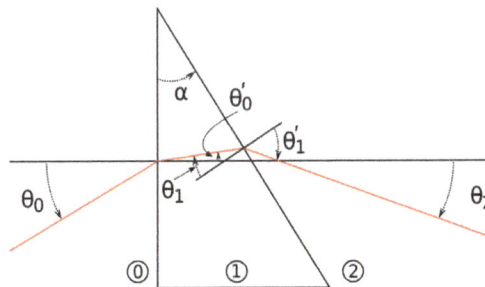

A ray trace through a prism with apex angle α. Regions 0, 1, and 2 have indices of refraction , , and , and primed angles indicate the ray's angle after refraction.

Ray angle deviation and dispersion through a prism can be determined by tracing a sample ray through the element and using Snell's law at each interface. For the prism shown at right, the indicated angles are given by

$$\theta_0' = \arcsin\left(\frac{n_0}{n_1}\sin\theta_0\right)$$

$$\theta_1 = \alpha - \theta_0'$$

$$\theta_1' = \arcsin\left(\frac{n_1}{n_2}\sin\theta_1\right)$$

$$\theta_2 = \theta_1' - \alpha$$

All angles are positive in the direction shown in the image. For a prism in air $n_0 = n_2 \approx 1$. Defining $n = n_1$, the deviation angle δ is given by

$$\delta = \theta_0 + \theta_2 = \theta_0 + \arcsin\left(n\sin\left[\alpha - \arcsin\left(\frac{1}{n}\sin\theta_0\right)\right]\right) - \alpha$$

If the angle of incidence θ_0 and prism apex angle α are both small, $\sin\theta \approx \theta$ and $\arcsin x \approx x$ if the angles are expressed in radians. This allows the nonlinear equation in the deviation angle δ to be approximated by

$$\delta \approx \theta_0 - \alpha + \left(n\left[\left(\alpha - \frac{1}{n}\theta_0\right)\right]\right) = \theta_0 - \alpha + n\alpha - \theta_0 = (n-1)\alpha.$$

The deviation angle depends on wavelength through n, so for a thin prism the deviation angle varies with wavelength according to

$$\delta(\lambda) \approx [n(\lambda) - 1]\alpha.$$

Prisms and the Nature of Light

Before Isaac Newton, it was believed that white light was colorless, and that the prism itself produced the color. Newton's experiments demonstrated that all the colors already existed in the light in a heterogeneous fashion, and that "corpuscles" (particles) of light were fanned out because particles with different colors traveled with different speeds through the prism. It was only later that Young and Fresnel combined Newton's particle theory with Huygens' wave theory to show that color is the visible manifestation of light's wavelength.

Newton arrived at his conclusion by passing the red color from one prism through a second prism and found the color unchanged. From this, he concluded that the colors must already be present in the incoming light — thus, the prism did not create colors, but merely separated colors that are already there. He also used a lens and a second prism to recompose the spectrum back into white light. This experiment has become a classic example of the methodology introduced during the scientific revolution. The results of this experiment dramatically transformed the field of metaphysics, leading to John Locke's primary vs secondary quality distinction.

Newton discussed prism dispersion in great detail in his book *Opticks*. He also introduced the use of more than one prism to control dispersion. Newton's description of his experiments on prism dispersion was qualitative, and is quite readable. A quantitative description of multiple-prism dispersion was not needed until multiple prism laser beam expanders were introduced in the 1980s.

Types of Prisms

Dispersive Prisms

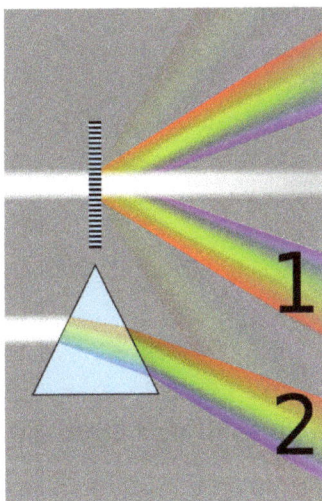

Comparison of the spectra obtained from a diffraction grating by diffraction (1), and a prism by refraction (2). Longer wavelengths (red) are diffracted more, but refracted less than shorter wavelengths (violet).

Dispersive prisms are used to break up light into its constituent spectral colors because the refractive index depends on frequency; the white light entering the prism is a mixture of different frequencies, each of which gets bent slightly differently. Blue light is slowed down more than red light and will therefore be bent more than red light.

- Triangular prism
- Abbe prism
- Pellin–Broca prism
- Amici prism
- Compound prism
- Grism, a dispersive prism with a diffraction grating on its surface

Reflective Prisms

Reflective prisms are used to reflect light, in order to flip, invert, rotate, deviate or displace the light beam. They are typically used to erect the image in binoculars or single-lens reflex cameras – without the prisms the image would be upside down for the user. Many reflective prisms use total internal reflection to achieve high reflectivity.

The most common reflective prisms are:

- Porro prism
- Porro–Abbe prism
- Amici roof prism
- Pentaprism and roof pentaprism
- Abbe–Koenig prism
- Schmidt–Pechan prism
- Bauernfeind prism
- Dove prism
- Retroreflector prism

Beam-splitting Prisms

Some reflective prisms are used for splitting a beam into two or more beams:

- Beam splitter cube
- Dichroic prism

Polarizing Prisms

There are also *polarizing prisms* which can split a beam of light into components of varying polarization. These are typically made of a birefringent crystalline material.

- Nicol prism
- Wollaston prism
- Nomarski prism – a variant of the Wollaston prism with advantages in microscopy
- Rochon prism
- Sénarmont prism

- Glan–Foucault prism
- Glan–Taylor prism
- Glan–Thompson prism

Deflecting Prisms

Wedge prisms are used to deflect a beam of light by a fixed angle. A pair of such prisms can be used for beam steering; by rotating the prisms the beam can be deflected into any desired angle within a conical "field of regard". The most commonly found implementation is a Risley prism pair. Two wedge prisms can also be used as an *anamorphic pair* to change the shape of a beam. This is used to make a round beam from the elliptical output of a laser diode.

Rhomboid prisms are used to laterally displace a beam of light without inverting the image.

Deck prisms were used on sailing ships to bring daylight below deck, since candles and kerosene lamps are a fire hazard on wooden ships.

In Optometry

By shifting corrective lenses off axis, images seen through them can be displaced in the same way that a prism displaces images. Eye care professionals use prisms, as well as lenses off axis, to treat various orthoptics problems:

- Diplopia (double vision)
- Positive and negative fusion problems
- Positive relative accommodation and negative relative accommodation problems.

Prism spectacles with a single prism perform a relative displacement of the two eyes, thereby correcting eso-, exo, hyper- or hypotropia.

In contrast, spectacles with prisms of equal power for both eyes, called yoked prisms (also: *conjugate prisms, ambient lenses* or *performance glasses*) shift the visual field of both eyes to the same extent.

Microscope

A microscope is an instrument used to see objects that are too small for the naked eye. The science of investigating small objects using such an instrument is called microscopy. Microscopic means invisible to the eye unless aided by a microscope.

There are many types of microscopes. The most common (and the first to be invented) is the optical microscope, which uses light to image the sample. Other major types of microscopes are the electron microscope (both the transmission electron microscope and the scanning electron microscope), the ultramicroscope, and the various types of scanning probe microscope.

18th century microscopes from the Musée des Arts et Métiers, Paris

History

The first microscope to be developed was the optical microscope, although the original inventor is not easy to identify. Evidence points to the first compound microscope appearing in the Netherlands by the 1620s, with a likely inventor being Cornelis Drebbel. Counter claims include it being invented by Hans Lippershey (who obtained the first telescope patent) and what may be a dubious claim by Zacharias Janssen's son that his father invented the microscope and telescope. Giovanni Faber coined the name microscope for Galileo Galilei's compound microscope in 1625 (Galileo had called it the "*occhiolino*" or "*little eye*").

Rise of Modern Light Microscopy

The first detailed account of the interior construction of living tissue based on the use of a microscope did not appear until 1644, in Giambattista Odierna's *L'occhio della mosca*, or *The Fly's Eye*.

It was not until the 1660s and 1670s that the microscope was used extensively for research in Italy, the Netherlands and England. Marcelo Malpighi in Italy began the analysis of biological structures beginning with the lungs. Robert Hooke's *Micrographia* had a huge impact, largely because of its impressive illustrations. The greatest contribution came from Antonie van Leeuwenhoek who achieved up to 300 times magnification. He sandwiched a v. small glass ball lens between the holes in two metal plates riveted together and with an adjustable by screws needle attached to mount the specimen. Then, Van

Leeuwenhoek discovered red blood cells and spermatozoa and helped popularise micros-copy as a technique. On 9 October 1676, he reported the discovery of micro-organisms.

The performance of light microscopy depends as much on how the sample is illuminat-ed as on how it is observed. Early instruments were limited until this principle was fully appreciated and developed, and until electric lamps were available as light sources. The first piece of fiction to involve the microcosm was probably Fitz-James O'Brien's "The Diamond Lens," which tells the story of a scientist who invents a powerful micro-scope and discovers a beautiful woman living in a microscopic world inside a drop of water. In 1893 August Köhler developed a key principle of sample illumination, Köhler illumination, which is central to achieving the theoretical limits of light microscopy. This method of sample illumination produces even lighting and overcomes the limited contrast and resolution imposed by early techniques of sample illumination. Further developments in sample illumination came from the discovery of phase contrast by Frits Zernike in 1953, and differential interference contrast illumination by Georges Nomarski in 1955; both of which allow imaging of unstained, transparent samples.

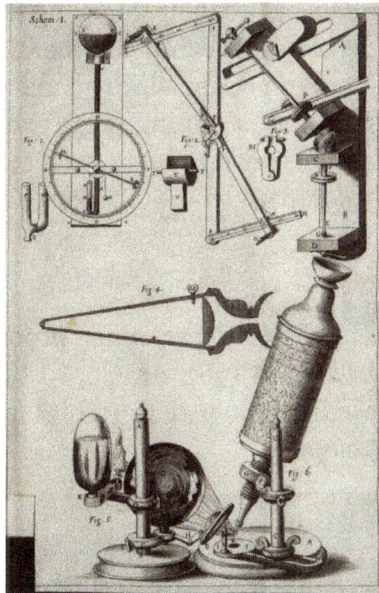

Robert Hooke's microscope and other Instruments

Electron Microscopy

In the early 20th century a significant alternative to light microscopy was devel-oped, using electrons rather than light to generate the image. Ernst Ruska started development of the first electron microscope in 1931 which was the transmission electron microscope (TEM). The transmission electron microscope works on the same principle as an optical microscope but uses electrons in the place of light and electromagnets in the place of glass lenses. Use of electrons instead of light allows a much higher resolution.

An ant as imaged using a scanning electron microscope (SEM)

Development of the transmission electron microscope was quickly followed in 1935 by the development of the scanning electron microscope by Max Knoll.

Electron microscopes quickly became popular following the Second World War. Ernst Ruska, working at Siemens, developed the first commercial transmission electron microscope and major scientific conferences on electron microscopy started being held in the 1950s. In 1965 the first commercial scanning electron microscope was developed by Professor Sir Charles Oatley and his postgraduate student Gary Stewart and marketed by the Cambridge Instrument Company as the "Stereoscan".

Scanning Probe Microscopy

The 1980s saw the development of the first scanning probe microscopes. The first was the scanning tunneling microscope in 1981, developed by Gerd Binnig and Heinrich Rohrer. This was closely followed in 1986 with Gerd Binnig, Quate, and Gerber's invention of the atomic force microscope.

Fluorescence and Light Microscopy

The most recent developments in light microscope largely centre on the rise of fluorescence microscopy in biology. During the last decades of the 20th century, particularly in the post-genomic era, many techniques for fluorescent labelling of cellular structures were developed. The main groups of techniques are small chemical staining of cellular structures, for example DAPI to label DNA, use of antibodies conjugated to fluorescent reporters, immunofluorescence, and fluorescent proteins, such as green fluorescent protein. These techniques use these different fluorophores for analysis of cell structure at a molecular level in both live and fixed samples.

The rise of fluorescence microscopy drove the development of a major modern microscope design, the confocal microscope. The principle was patented in 1957 by Marvin Minsky, although laser technology limited practical application of the technique. It was not until 1978 when Thomas and Christoph Cremer developed the first practical confocal laser scanning microscope and the technique rapidly gained popularity through the 1980s.

Much current research (in the early 21st century) on optical microscope techniques is focused on development of superresolution analysis of fluorescently labelled samples. Structured illumination can improve resolution by around two to four times and techniques like stimulated Emission Depletion microscopy are approaching the resolution of electron microscopes.

Types

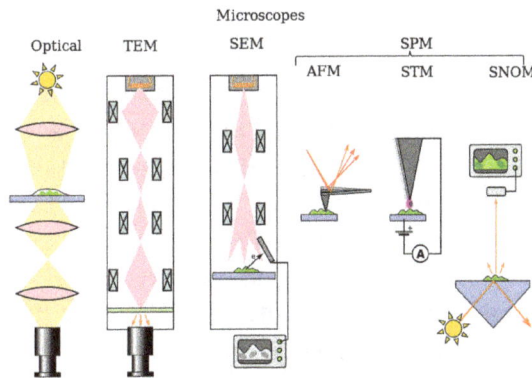

Types of microscopes

Microscopes can be separated into several different classes. One grouping is based on what interacts with the sample to generate the image, i.e., light or photons (optical microscopes), electrons (electron microscopes) or a probe (scanning probe microscopes). Alternatively, microscopes can be classed on whether they analyze the sample via a scanning point (confocal optical microscopes, scanning electron microscopes and scanning probe microscopes) or analyze the sample all at once (wide field optical microscope and transmission electron microscopes).

Wide field optical microscopes and transmission electron microscopes both use the theory of lenses (optics for light microscopes and electromagnet lenses for electron microscopes) in order to magnify the image generated by the passage of a wave transmitted through the sample, or reflected by the sample. The waves used are electromagnetic (in optical microscopes) or electron beams (in electron microscopes). Resolution in these microscopes is limited by the wavelength of the radiation used to image the sample, where shorter wavelengths allow for a higher resolution.

Scanning optical and electron microscopes, like the confocal microscope and scanning electron microscope, use lenses to focus a spot of light or electrons onto the sample then analyze the reflected or transmitted waves. The point is then scanned over the sample to analyze a rectangular region. Magnification of the image is achieved by displaying the data from scanning a physically small sample area on a relatively large screen. These microscopes have the same resolution limit as wide field optical, probe, and electron microscopes.

Scanning probe microscopes also analyze a single point in the sample and then scan the probe over a rectangular sample region to build up an image. As these microscopes do not use electromagnetic or electron radiation for imaging they are not subject to the same resolution limit as the optical and electron microscopes described above.

Optical

The most common type of microscope (and the first invented) is the optical microscope. This is an optical instrument containing one or more lenses producing an enlarged image of a sample placed in the focal plane. Optical microscopes have refractive glass and occasionally of plastic or quartz, to focus light into the eye or another light detector. Mirror-based optical microscopes operate in the same manner. Typical magnification of a light microscope, assuming visible range light, is up to 1250x with a theoretical resolution limit of around 0.250 micrometres or 250 nanometres. This limits the practical magnification limit to ~1500x. Specialized techniques (e.g., scanning confocal microscopy, Vertico SMI) may exceed this magnification but the resolution is diffraction limited. The use of shorter wavelengths of light, such as the ultraviolet, is one way to improve the spatial resolution of the optical microscope, as are devices such as the near-field scanning optical microscope. Sarfus, a recent optical technique increases the sensitivity of standard optical microscope to a point it becomes possible to directly visualize nanometric films (down to 0.3 nanometre) and isolated nano-objects (down to 2 nm-diameter). The technique is based on the use of non-reflecting substrates for cross-polarized reflected light microscopy.

Ultraviolet light enables the resolution of microscopic features, as well as to image samples that are transparent to the eye. Near infrared light can be used to visualize circuitry embedded in bonded silicon devices, since silicon is transparent in this region of wavelengths.

In fluorescence microscopy, many wavelengths of light, ranging from the ultraviolet to the visible can be used to cause samples to fluoresce to allow viewing by eye or with the use of specifically sensitive cameras.

Phase contrast microscopy is an optical microscopy illumination technique in which small phase shifts in the light passing through a transparent specimen are converted into amplitude or contrast changes in the image. The use of phase contrast does not require staining to view the slide. This microscope technique made it possible to study the cell cycle in live cells.

The traditional optical microscope has more recently evolved into the digital microscope. In addition to, or instead of, directly viewing the object through the eyepieces, a type of sensor similar to those used in a digital camera is used to obtain an image, which is then displayed on a computer monitor. These sensors may use CMOS or charge-coupled device (CCD) technology, depending on the application.

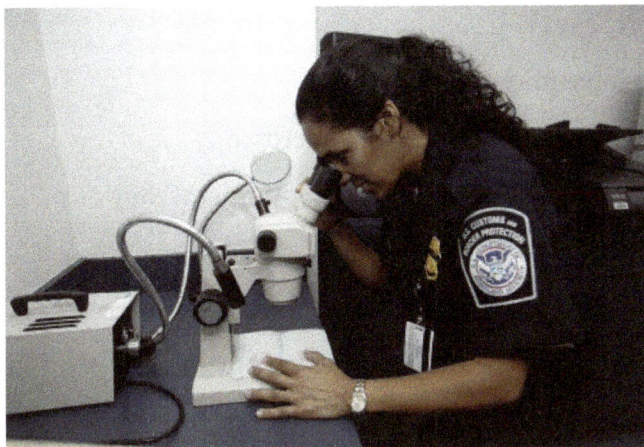

CBP Office of Field Operations agent checking the authenticity of a travel document at an international airport using a stereo microscope

Electron

Three major types of electron microscopes exist:

- Scanning electron microscope (SEM): looks at the surface of bulk objects by scanning the surface with a fine electron beam.

- Transmission electron microscope (TEM): passes electrons through the sample, analogous to basic optical microscopy. This requires careful sample preparation, since electrons are scattered so strongly by most materials. This is a scientific device that allows people to see objects that could normally not be seen by the naked or unaided eye.

- Hybrid instruments. Combination of both types.

Scanning Probe

- AFM, atomic force microscopy

- BEEM, ballistic electron emission microscopy

- EFM, electrostatic force microscope

- ESTM electrochemical scanning tunneling microscope

- FMM, force modulation microscopy

- KPFM, kelvin probe force microscopy

- MFM, magnetic force microscopy

- MRFM, magnetic resonance force microscopy

- NSOM, near-field scanning optical microscopy (or SNOM, scanning near-field optical microscopy)

- PFM, piezo force microscopy

- PSTM, photon scanning tunneling microscopy

- PTMS, photothermal microspectroscopy/microscopy

- SAP, scanning atom probe

- SCM, scanning capacitance microscopy

- SECM, scanning electrochemical microscopy

- SGM, scanning gate microscopy

- SICM, scanning ion-conductance microscopy

- SPSM spin polarized scanning tunneling microscopy

- SThM, scanning thermal microscopy

- STM, scanning tunneling microscopy

- SVM, scanning voltage microscopy

- SHPM, scanning Hall probe microscopy

- SSM, Scanning SQUID microscope

- RTM, Recurrence tracking microscope

Of these techniques AFM and STM are the most commonly used.

Other Types

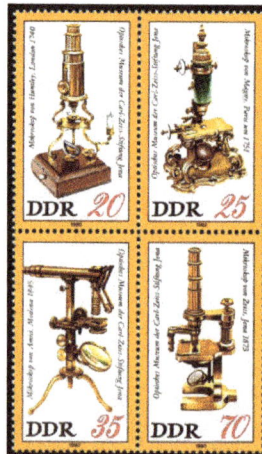

Different microscopes

Scanning acoustic microscopes use sound waves to measure variations in acoustic impedance. Similar to Sonar in principle, they are used for such jobs as detecting defects in the subsurfaces of materials including those found in integrated circuits. On February 4, 2013, Australian engineers built a "quantum microscope" which provides unparalleled precision.

Optical Microscope

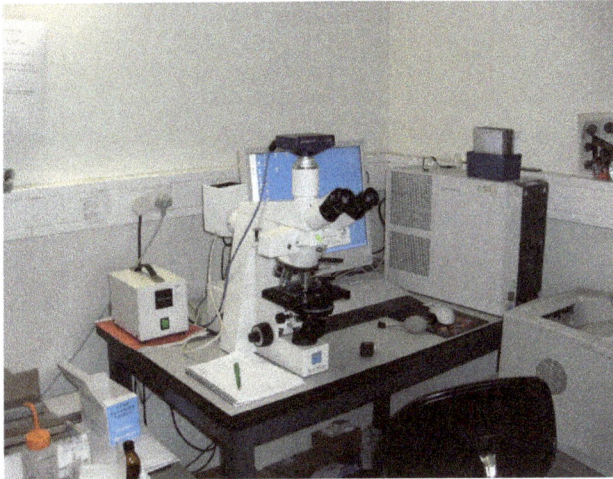

A modern microscope with a mercury bulb for fluorescence microscopy. The microscope has a digital camera, and is attached to a computer.

The optical microscope, often referred to as light microscope, is a type of microscope which uses visible light and a system of lenses to magnify images of small samples. Optical microscopes are the oldest design of microscope and were possibly invented in their present compound form in the 17th century. Basic optical microscopes can be very simple, although there are many complex designs which aim to improve resolution and sample contrast.

The image from an optical microscope can be captured by normal light-sensitive cameras to generate a micrograph. Originally images were captured by photographic film but modern developments in CMOS and charge-coupled device (CCD) cameras allow the capture of digital images. Purely digital microscopes are now available which use a CCD camera to examine a sample, showing the resulting image directly on a computer screen without the need for eyepieces.

Alternatives to optical microscopy which do not use visible light include scanning electron microscopy and transmission electron microscopy.

On 8 October 2014, the Nobel Prize in Chemistry was awarded to Eric Betzig, William Moerner and Stefan Hell for "the development of super-resolved fluorescence microscopy," which brings "optical microscopy into the nanodimension".

Optical Configurations

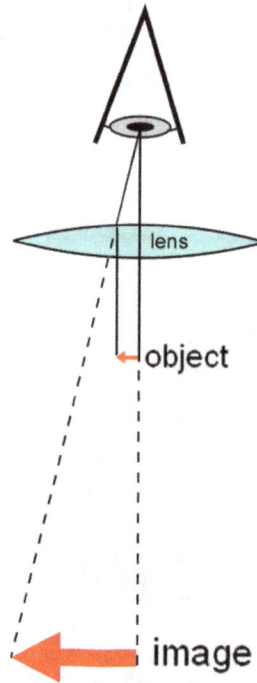

Diagram of a simple microscope

There are two basic configurations of the conventional optical microscope: the simple microscope and the compound microscope. The vast majority of modern research microscopes are compound microscopes while some cheaper commercial digital microscopes are simple single lens microscopes. A magnifying glass is, in essence, a single lens simple microscope. In general, microscope optics are static; to focus at different focal depths the lens to sample distance is adjusted, and to get a wider or narrower field of view a different magnification objective lens must be used. Most modern research microscopes also have a separate set of optics for illuminating the sample.

Simple Microscope

A simple microscope is a microscope that uses a lens or set of lenses to enlarge an object through angular magnification alone, giving the viewer an erect enlarged virtual image. Simple microscopes are not capable of high magnification. The use of a single convex lens or groups of lenses are still found in simple magnification devices such as the magnifying glass, loupes, and eyepieces for telescopes and microscopes.

Compound Microscope

A compound microscope is a microscope which uses a lens close to the object being viewed to collect light (called the objective lens) which focuses a real image of the object inside the microscope (image 1). That image is then magnified by a second lens or group of lenses (called the eyepiece) that gives the viewer an enlarged inverted virtual image of the object (image 2). The use of a compound objective/eyepiece combination allows for much higher magnification, reduced chromatic aberration and exchangeable objective lenses to adjust the magnification. A compound microscope also enables more advanced illumination setups, such as phase contrast.

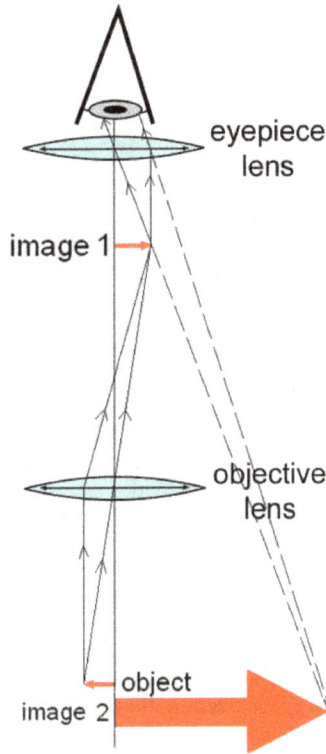

Diagram of a compound microscope

History

Invention

It is difficult to say who invented the compound microscope. The Dutch spectacle-maker Zacharias Janssen is sometimes claimed to have invented it in 1590 (a claim made by his son and fellow countrymen, in different testimony in 1634 and 1655). Another claim is that Janssen's competitor, Hans Lippershey, invented the compound microscope. Another favorite for the title of 'inventor of the microscope' was Galileo Galilei. He developed an *occhiolino* or compound microscope with a convex and a concave lens in 1609. Galileo's microscope was celebrated in the Accademia dei Lincei in 1624 and was the first such device to be given the name "microscope" a year later by fellow Lincean Giovanni Faber.

Christiaan Huygens, another Dutchman, developed a simple 2-lens ocular system in the late 17th century that was achromatically corrected, and therefore a huge step forward in microscope development. The Huygens ocular is still being produced to this day, but suffers from a small field size, and other minor disadvantages.

Popularization

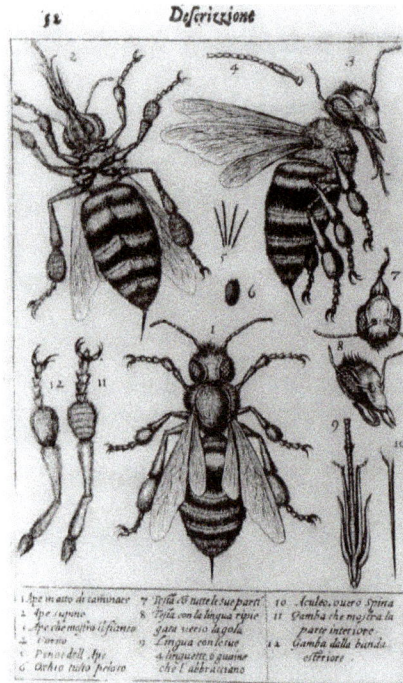

The oldest published image known to have been made with a microscope: bees by Francesco Stelluti, 1630

Antonie van Leeuwenhoek (1632–1724) is credited with bringing the microscope to the attention of biologists, even though simple magnifying lenses were already being produced in the 16th century. Van Leeuwenhoek's home-made microscopes were simple microscopes, with a single very small, yet strong lens. They were awkward in use, but enabled van Leeuwenhoek to see detailed images. It took about 150 years of optical development before the compound microscope was able to provide the same quality image as van Leeuwenhoek's simple microscopes, due to difficulties in configuring multiple lenses.

Lighting Techniques

While basic microscope technology and optics have been available for over 400 years it is much more recently that techniques in sample illumination were developed to generate the high quality images seen today.

In August 1893 August Köhler developed Köhler illumination. This method of sample illumination gives rise to extremely even lighting and overcomes many limitations of older techniques of sample illumination. Before development of Köhler illumination the image of the light source, for example a lightbulb filament, was always visible in the image of the sample.

The Nobel Prize in physics was awarded to Dutch physicist Frits Zernike in 1953 for his development of phase contrast illumination which allows imaging of transparent samples. By using interference rather than absorption of light, extremely transparent samples, such as live mammalian cells, can be imaged without having to use staining techniques. Just two years later, in 1955, Georges Nomarski published the theory for differential interference contrast microscopy, another interference-based imaging technique.

Fluorescence Microscopy

Modern biological microscopy depends heavily on the development of fluorescent probes for specific structures within a cell. In contrast to normal transilluminated light microscopy, in fluorescence microscopy the sample is illuminated through the objective lens with a narrow set of wavelengths of light. This light interacts with fluorophores in the sample which then emit light of a longer wavelength. It is this emitted light which makes up the image.

Since the mid 20th century chemical fluorescent stains, such as DAPI which binds to DNA, have been used to label specific structures within the cell. More recent developments include immunofluorescence, which uses fluorescently labelled antibodies to recognise specific proteins within a sample, and fluorescent proteins like GFP which a live cell can express making it fluorescent.

Components

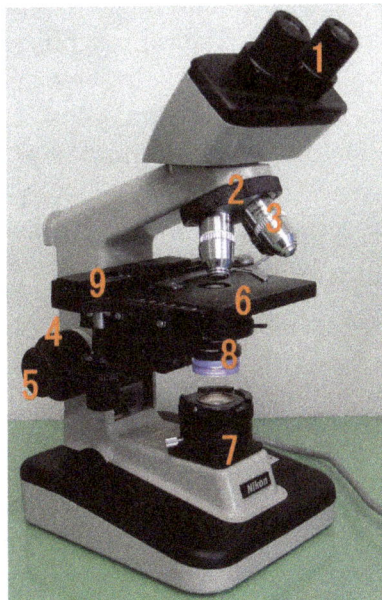

Basic optical transmission microscope elements (1990s)

All modern optical microscopes designed for viewing samples by transmitted light share the same basic components of the light path. In addition, the vast majority of microscopes have the same 'structural' components (numbered below according to the image on the right):

- Eyepiece (ocular lens) (1)

- Objective turret, revolver, or revolving nose piece (to hold multiple objective lenses) (2)

- Objective lenses (3)

- Focus knobs (to move the stage)

    o   Coarse adjustment (4)

    o   Fine adjustment (5)

- Stage (to hold the specimen) (6)

- Light source (a light or a mirror) (7)

- Diaphragm and condenser (8)

- Mechanical stage (9)

Eyepiece (Ocular Lens)

The eyepiece, or ocular lens, is a cylinder containing two or more lenses; its function is to bring the image into focus for the eye. The eyepiece is inserted into the top end of the body tube. Eyepieces are interchangeable and many different eyepieces can be inserted with different degrees of magnification. Typical magnification values for eyepieces include 5×, 10× (the most common), 15X and 20×. In some high performance microscopes, the optical configuration of the objective lens and eyepiece are matched to give the best possible optical performance. This occurs most commonly with apochromatic objectives.

Objective Turret (Revolver or Revolving Nose Piece)

Objective turret, revolver, or revolving nose piece is the part that holds the set of objective lenses. It allows the user to switch between objective lenses.

Objective

At the lower end of a typical compound optical microscope, there are one or more objective lenses that collect light from the sample. The objective is usually in a cylinder housing containing a glass single or multi-element compound lens. Typically there will be around three objective lenses screwed into a circular nose piece which may be rotated to select the required objective lens. These arrangements are designed to be parfocal, which means that when one changes from one lens to another on a microscope, the sample stays in focus. Microscope objectives are characterized by two parameters, namely, magnification and numerical aperture. The former typically ranges from 5× to 100× while the latter ranges from 0.14 to 0.7, corresponding to focal lengths of about 40 to 2 mm, respectively. Objective lenses with higher magnifications normally have a higher numerical aperture and a shorter depth of field in the resulting image. Some high performance objective lenses may require matched eyepieces to deliver the best optical performance.

Oil Immersion Objective

Two Leica oil immersion microscope objective lenses: 100× (left) and 40× (right)

Some microscopes make use of oil-immersion objectives or water-immersion objectives for greater resolution at high magnification. These are used with index-matching material such as immersion oil or water and a matched cover slip between the objective lens and the sample. The refractive index of the index-matching material is higher than air allowing the objective lens to have a larger numerical aperture (greater than 1) so that the light is transmitted from the specimen to the outer face of the objective lens with minimal refraction. Numerical apertures as high as 1.6 can be achieved. The larger numerical aperture allows collection of more light making detailed observation of smaller details possible. An oil immersion lens usually has a magnification of 40 to 100×.

Focus Knobs

Adjustment knobs move the stage up and down with separate adjustment for coarse and fine focusing. The same controls enable the microscope to adjust to specimens of different thickness. In older designs of microscopes, the focus adjustment wheels move the microscope tube up or down relative to the stand and had a fixed stage.

Frame

The whole of the optical assembly is traditionally attached to a rigid arm, which in turn is attached to a robust U-shaped foot to provide the necessary rigidity. The arm angle may be adjustable to allow the viewing angle to be adjusted.

The frame provides a mounting point for various microscope controls. Normally this will include controls for focusing, typically a large knurled wheel to adjust coarse focus, together with a smaller knurled wheel to control fine focus. Other features may be lamp controls and/or controls for adjusting the condenser.

Stage

The stage is a platform below the objective which supports the specimen being viewed. In the center of the stage is a hole through which light passes to illuminate the specimen. The stage usually has arms to hold slides (rectangular glass plates with typical dimensions of 25×75 mm, on which the specimen is mounted).

At magnifications higher than 100× moving a slide by hand is not practical. A mechanical stage, typical of medium and higher priced microscopes, allows tiny movements of the slide via control knobs that reposition the sample/slide as desired. If a microscope did not originally have a mechanical stage it may be possible to add one.

All stages move up and down for focus. With a mechanical stage slides move on two horizontal axes for positioning the specimen to examine specimen details.

Focusing starts at lower magnification in order to center the specimen by the user on the stage. Moving to a higher magnification requires the stage to be moved higher vertically for re-focus at the higher magnification and may also require slight horizontal specimen position adjustment. Horizontal specimen position adjustments are the reason for having a mechanical stage.

Due to the difficulty in preparing specimens and mounting them on slides, for children it's best to begin with prepared slides that are centered and focus easily regardless of the focus level used.

Light Source

Many sources of light can be used. At its simplest, daylight is directed via a mirror. Most microscopes, however, have their own adjustable and controllable light source – often a halogen lamp, although illumination using LEDs and lasers are becoming a more common provision.

Condenser

The condenser is a lens designed to focus light from the illumination source onto the sample. The condenser may also include other features, such as a diaphragm and/or filters, to manage the quality and intensity of the illumination. For illumination techniques like dark field, phase contrast and differential interference contrast microscopy additional optical components must be precisely aligned in the light path.

Magnification

The actual power or magnification of a compound optical microscope is the product of the powers of the ocular (eyepiece) and the objective lens. The maximum normal magnifications of the ocular and objective are 10× and 100× respectively, giving a final magnification of 1,000×.

Magnification and Micrographs

When using a camera to capture a micrograph the effective magnification of the image must take into account the size of the image. This is independent of whether it is on a print from a film negative or displayed digitally on a computer screen.

In the case of photographic film cameras the calculation is simple; the final magnification is the product of: the objective lens magnification, the camera optics magnification and the enlargement factor of the film print relative to the negative. A typical value of the enlargement factor is around 5× (for the case of 35mm film and a 15x10 cm (6×4 inch) print).

In the case of digital cameras the size of the pixels in the CMOS or CCD detector and the size of the pixels on the screen have to be known. The enlargement factor from the detector to the pixels on screen can then be calculated. As with a film camera the final magnification is the product of: the objective lens magnification, the camera optics magnification and the enlargement factor.

Operation

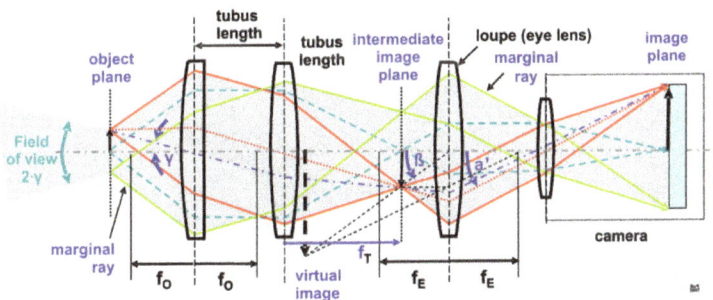

Optical path in a typical microscope

The optical components of a modern microscope are very complex and for a microscope to work well, the whole optical path has to be very accurately set up and controlled. Despite this, the basic operating principles of a microscope are quite simple.

The objective lens is, at its simplest, a very high powered magnifying glass *i.e.* a lens with a very short focal length. This is brought very close to the specimen being examined so that the light from the specimen comes to a focus about 160 mm inside the microscope tube. This creates an enlarged image of the subject. This image is inverted and can be seen by removing the eyepiece and placing a piece of tracing paper over the end of the tube. By carefully focusing a brightly lit specimen, a highly enlarged image can be seen. It is this real image that is viewed by the eyepiece lens that provides further enlargement.

In most microscopes, the eyepiece is a compound lens, with one component lens near the front and one near the back of the eyepiece tube. This forms an air-separated couplet. In many designs, the virtual image comes to a focus between the two lenses of the eyepiece, the first lens bringing the real image to a focus and the second lens enabling the eye to focus on the virtual image.

In all microscopes the image is intended to be viewed with the eyes focused at infinity (mind that the position of the eye in the above figure is determined by the eye's focus). Headaches and tired eyes after using a microscope are usually signs that the eye is being forced to focus at a close distance rather than at infinity.

Illumination Techniques

Many techniques are available which modify the light path to generate an improved contrast image from a sample. Major techniques for generating increased contrast from the sample include cross-polarized light, dark field, phase contrast and differential interference contrast illumination. A recent technique (Sarfus) combines cross-polarized light and specific contrast-enhanced slides for the visualization of nanometric samples.

- Four examples of transilumination techniques used to generate contrast in a sample of tissue paper. 1.559 µm/pixel.

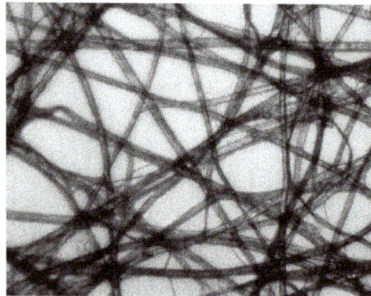

Bright field illumination, sample contrast comes from absorbance of light in the sample.

Cross-polarized light illumination, sample contrast comes from rotation of polarized light through the sample.

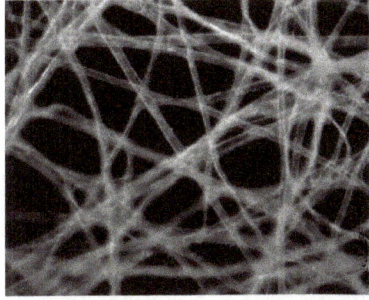

Dark field illumination, sample contrast comes from light scattered by the sample.

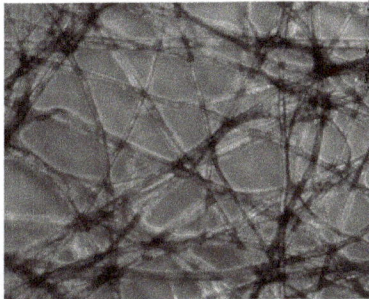

Phase contrast illumination, sample contrast comes from interference of different path lengths of light through the sample.

Other Techniques

Modern microscopes allow more than just observation of transmitted light image of a sample; there are many techniques which can be used to extract other kinds of data. Most of these require additional equipment in addition to a basic compound microscope.

- Reflected light, or incident, illumination (for analysis of surface structures)

- Fluorescence microscopy, both:

  - Epifluorescence microscopy

  - Confocal microscopy

- Microspectroscopy (where a UV-visible spectrophotometer is integrated with an optical microscope)

- Ultraviolet microscopy

- Near-Infrared microscopy

- Multiple transmission microscopy for contrast enhancement and aberration reduction.

- Automation (for automatic scanning of a large sample or image capture)

Applications

Optical microscopy is used extensively in microelectronics, nanophysics, biotechnology, pharmaceutic research, mineralogy and microbiology.

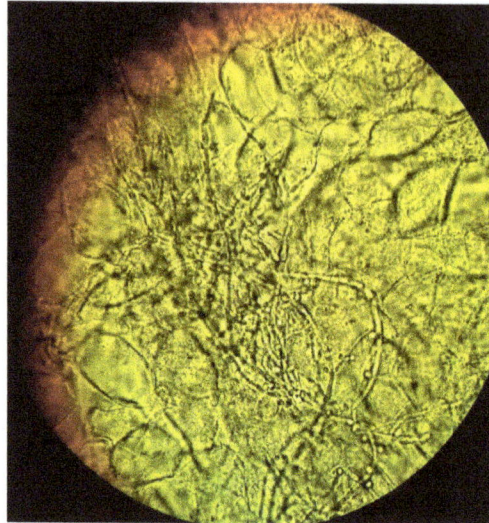

A 40x magnification image of cells in a medical smear test taken through an optical microscope using a wet mount technique, placing the specimen on a glass slide and mixing with a salt solution

Optical microscopy is used for medical diagnosis, the field being termed histopathology when dealing with tissues, or in smear tests on free cells or tissue fragments.

In industrial use, binocular microscopes are common. Aside from applications needing true depth perception, the use of dual eyepieces reduces eye strain associated with long workdays at a microscopy station. In certain applications, long-working-distance or long-focus microscopes are beneficial. An item may need to be examined behind a window, or industrial subjects may be a hazard to the objective. Such optics resemble telescopes with close-focus capabilities.

Optical Microscope Variants

There are many variants of the basic compound optical microscope design for specialized purposes. Some of these are physical design differences allowing specialization for certain purposes:

- Stereo microscope, a low powered microscope which provides a stereoscopic view of the sample, commonly used for dissection.

- Comparison microscope, which has two separate light paths allowing direct comparison of two samples via one image in each eye.

- Inverted microscope, for studying samples from below; useful for cell cultures in liquid, or for metallography.

- Student microscope, designed for low cost, durability, and ease of use.

- Fiber optic connector inspection microscope, designed for connector end-face inspection

Other microscope variants are designed for different illumination techniques:

- Petrographic microscope, whose design usually includes a polarizing filter, rotating stage and gypsum plate to facilitate the study of minerals or other crystalline materials whose optical properties can vary with orientation.

- Polarizing microscope, similar to the petrographic microscope.

- Phase contrast microscope, which applies the phase contrast illumination method.

- Epifluorescence microscope, designed for analysis of samples which include fluorophores.

- Confocal microscope, a widely used variant of epifluorescent illumination which uses a scanning laser to illuminate a sample for fluorescence.

- Student microscope – an often low-power microscope with simplified controls and sometimes low quality optics designed for school use or as a starter instrument for children.

Digital Microscope

A digital microscope is a microscope equipped with a digital camera allowing observation of a sample via a computer. Microscopes can also be partly or wholly computer-controlled with various levels of automation. Digital microscopy allows greater analysis of a microscope image, for example measurements of distances and areas and quantitaton of a fluorescent or histological stain.

A miniature USB microscope.

Low-powered digital microscopes, USB microscopes, are also commercially available.

These are essentially webcams with a high-powered macro lens and generally do not use transillumination. The camera attached directly to the USB port of a computer, so that the images are shown directly on the monitor. They offer modest magnifications (up to about 200×) without the need to use eyepieces, and at very low cost. High power illumination is usually provided by an LED source or sources adjacent to the camera lens.

Limitations

At very high magnifications with transmitted light, point objects are seen as fuzzy discs surrounded by diffraction rings. These are called Airy disks. The *resolving power* of a microscope is taken as the ability to distinguish between two closely spaced Airy disks (or, in other words the ability of the microscope to reveal adjacent structural detail as distinct and separate). It is these impacts of diffraction that limit the ability to resolve fine details. The extent and magnitude of the diffraction patterns are affected by both the wavelength of light (λ), the refractive materials used to manufacture the objective lens and the numerical aperture (NA) of the objective lens. There is therefore a finite limit beyond which it is impossible to resolve separate points in the objective field, known as the diffraction limit. Assuming that optical aberrations in the whole optical set-up are negligible, the resolution d, can be stated as:

$$d = \frac{\lambda}{2NA}$$

Usually a wavelength of 550 nm is assumed, which corresponds to green light. With air as the external medium, the highest practical *NA* is 0.95, and with oil, up to 1.5. In practice the lowest value of d obtainable with conventional lenses is about 200 nm. A new type of lens using multiple scattering of light allowed to improve the resolution to below 100 nm.

Surpassing the Resolution Limit

Multiple techniques are available for reaching resolutions higher than the transmitted light limit described above. Holographic techniques, as described by Courjon and Bulabois in 1979, are also capable of breaking this resolution limit, although resolution was restricted in their experimental analysis.

Using fluorescent samples more techniques are available. Examples include Vertico SMI, near field scanning optical microscopy which uses evanescent waves, and stimulated emission depletion. In 2005, a microscope capable of detecting a single molecule was described as a teaching tool.

Despite significant progress in the last decade, techniques for surpassing the diffraction limit remain limited and specialized.

While most techniques focus on increases in lateral resolution there are also some techniques which aim to allow analysis of extremely thin samples. For example, sarfus methods place the thin sample on a contrast-enhancing surface and thereby allows to directly visualize films as thin as 0.3 nanometers.

Structured Illumination SMI

SMI (spatially modulated illumination microscopy) is a light optical process of the so-called point spread function (PSF) engineering. These are processes which modify the PSF of a microscope in a suitable manner to either increase the optical resolution, to maximize the precision of distance measurements of fluorescent objects that are small relative to the wavelength of the illuminating light, or to extract other structural parameters in the nanometer range.

Localization Microscopy SPDMphymod

SPDM (spectral precision distance microscopy), the basic localization microscopy technology is a light optical process of fluorescence microscopy which allows position, distance and angle measurements on "optically isolated" particles (e.g. molecules) well below the theoretical limit of resolution for light microscopy. "Optically isolated" means that at a given point in time, only a single particle/molecule within a region of a size determined by conventional optical resolution (typically approx. 200–250 nm diameter) is being registered. This is possible when molecules within such a region all carry different spectral markers (e.g. different colors or other usable differences in the light emission of different particles).

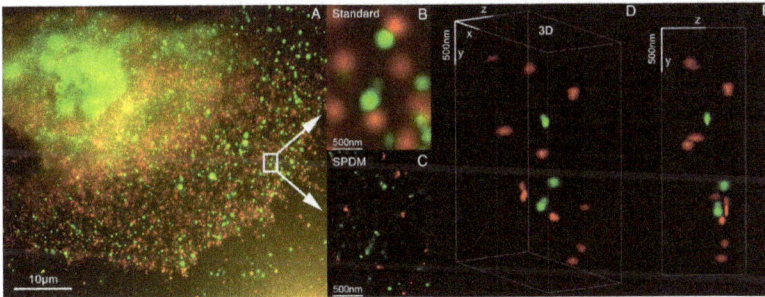

3D dual color super resolution microscopy with Her2 and Her3 in breast cells, standard dyes: Alexa 488, Alexa 568 LIMON

Many standard fluorescent dyes like GFP, Alexa dyes, Atto dyes, Cy2/Cy3 and fluorescein molecules can be used for localization microscopy, provided certain photo-physical conditions are present. Using this so-called SPDMphymod (physically modifiable fluorophores) technology a single laser wavelength of suitable intensity is sufficient for nanoimaging.

3D Super Resolution Microscopy

3D super resolution microscopy with standard fluorescent dyes can be achieved by combination of localization microscopy for standard fluorescent dyes SPDMphymod and structured illumination SMI.

STED

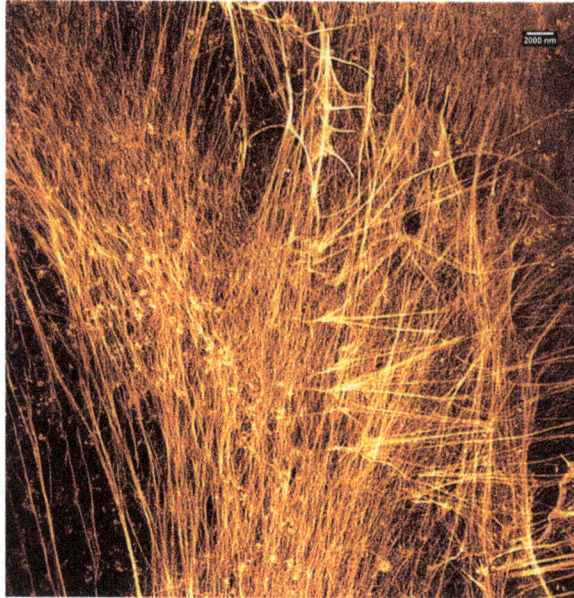

Stimulated emission depletion (STED) microscopy image of actin filaments within a cell.

Stimulated emission depletion is a simple example of how higher resolution surpassing the diffraction limit is possible, but it has major limitations. STED is a fluorescence microscopy technique which uses a combination of light pulses to induce fluorescence in a small sub-population of fluorescent molecules in a sample. Each molecule produces a diffraction-limited spot of light in the image, and the centre of each of these spots corresponds to the location of the molecule. As the number of fluorescing molecules is low the spots of light are unlikely to overlap and therefore can be placed accurately. This process is then repeated many times to generate the image. Stefan Hell of the Max Planck Institute for Biophysical Chemistry was awarded the 10th German Future Prize in 2006 and Nobel Prize for Chemistry in 2014 for his development of the STED microscope and associated methodologies.

Alternatives

In order to overcome the limitations set by the diffraction limit of visible light other microscopes have been designed which use other waves.

- Atomic force microscope (AFM)

- Scanning electron microscope (SEM)

- Scanning ion-conductance microscopy (SICM)

- Scanning tunneling microscope (STM)

- Transmission electron microscopy (TEM)

- Ultraviolet microscope

- X-ray microscope

It is important to note that higher frequency waves have limited interaction with matter, for example soft tissues are relatively transparent to X-rays resulting in distinct sources of contrast and different target applications.

The use of electrons and X-rays in place of light allows much higher resolution – the wavelength of the radiation is shorter so the diffraction limit is lower. To make the short-wavelength probe non-destructive, the atomic beam imaging system (atomic nanoscope) has been proposed and widely discussed in the literature, but it is not yet competitive with conventional imaging systems.

STM and AFM are scanning probe techniques using a small probe which is scanned over the sample surface. Resolution in these cases is limited by the size of the probe; micromachining techniques can produce probes with tip radii of 5–10 nm.

Additionally, methods such as electron or X-ray microscopy use a vacuum or partial vacuum, which limits their use for live and biological samples (with the exception of an environmental scanning electron microscope). The specimen chambers needed for all such instruments also limits sample size, and sample manipulation is more difficult. Color cannot be seen in images made by these methods, so some information is lost. They are however, essential when investigating molecular or atomic effects, such as age hardening in aluminium alloys, or the microstructure of polymers.

Telescope

A telescope is an optical instrument that aids in the observation of remote objects by collecting electromagnetic radiation (such as visible light). The first known practical telescopes were invented in the Netherlands at the beginning of the 1600s, by using glass lenses. They found use in both terrestrial applications and astronomy.

Within a few decades, the reflecting telescope was invented, which used mirrors to collect and focus the light. In the 20th century many new types of telescopes were invented, including radio telescopes in the 1930s and infrared telescopes in the 1960s. The word *telescope* now refers to a wide range of instruments capable of detecting different regions of the electromagnetic spectrum, and in some cases other types of detectors.

The word *telescope* was coined in 1611 by the Greek mathematician Giovanni Demisiani for one of Galileo Galilei's instruments presented at a banquet at the Accademia dei Lincei. In the *Starry Messenger*, Galileo had used the term *perspicillum*.

The 100 inch (2.54 m) Hooker reflecting telescope at Mount Wilson Observatory
near Los Angeles, USA.

History

Modern telescopes typically use CCDs instead of film for recording images. This is the sensor
array in the Kepler spacecraft.

28-inch telescope and 40-foot telescope in Greenwich in 2015.

The earliest recorded working telescopes were the refracting telescopes that appeared in the Netherlands in 1608. Their development is credited to three individuals: Hans Lippershey and Zacharias Janssen, who were spectacle makers in Middelburg, and Jacob Metius of Alkmaar. Galileo heard about the Dutch telescope in June 1609, built his own within a month, and improved upon the design in the following year. In the same year, Galileo became the first person to point a telescope skyward in order to make telescopic observations of a celestial object.

The idea that the objective, or light-gathering element, could be a mirror instead of a lens was being investigated soon after the invention of the refracting telescope. The potential advantages of using parabolic mirrors—reduction of spherical aberration and no chromatic aberration—led to many proposed designs and several attempts to build reflecting telescopes. In 1668, Isaac Newton built the first practical reflecting telescope, of a design which now bears his name, the Newtonian reflector.

The invention of the achromatic lens in 1733 partially corrected color aberrations present in the simple lens and enabled the construction of shorter, more functional refracting telescopes. Reflecting telescopes, though not limited by the color problems seen in refractors, were hampered by the use of fast tarnishing speculum metal mirrors employed during the 18th and early 19th century—a problem alleviated by the introduction of silver coated glass mirrors in 1857, and aluminized mirrors in 1932. The maximum physical size limit for refracting telescopes is about 1 meter (40 inches), dictating that the vast majority of large optical researching telescopes built since the turn of the 20th century have been reflectors. The largest reflecting telescopes currently have objectives larger than 10 m (33 feet), and work is underway on several 30-40m designs.

The 20th century also saw the development of telescopes that worked in a wide range of wavelengths from radio to gamma-rays. The first purpose built radio telescope went into operation in 1937. Since then, a tremendous variety of complex astronomical instruments have been developed.

Types

The name "telescope" covers a wide range of instruments. Most detect electromagnetic radiation, but there are major differences in how astronomers must go about collecting light (electromagnetic radiation) in different frequency bands.

Telescopes may be classified by the wavelengths of light they detect:

- X-ray telescopes, using shorter wavelengths than ultraviolet light
- Ultraviolet telescopes, using shorter wavelengths than visible light
- Optical telescopes, using visible light
- Infrared telescopes, using longer wavelengths than visible light
- Submillimetre telescopes, using longer wavelengths than infrared light
- Fresnel Imager, an optical lens technology
- X-ray optics, optics for certain X-ray wavelengths

Light Comparison						
Name	**Wavelength**	**Frequency (Hz)**	**Photon Energy (eV)**			
Gamma ray	less than 0.01 nm	more than 10 EHZ	100 keV – 300+ GeV	X		
X-Ray	0.01 to 10 nm	30 PHz – 30 EHZ	120 eV to 120 keV	X		
Ultraviolet	10 nm – 400 nm	30 EHZ – 790 THz	3 eV to 124 eV			
Visible	390 nm – 750 nm	790 THz – 405 THz	1.7 eV – 3.3 eV	X		
Infrared	750 nm – 1 mm	405 THz – 300 GHz	1.24 meV – 1.7 eV	X		
Microwave	1 mm – 1 meter	300 GHz – 300 MHz	1.24 meV – 1.24 µeV			
Radio	1 mm – km	300 GHz – 3 Hz	1.24 meV – 12.4 feV	X		

As wavelengths become longer, it becomes easier to use antenna technology to interact with electromagnetic radiation (although it is possible to make very tiny antenna). The near-infrared can be collected much like visible light, however in the far-infrared and submillimetre range, telescopes can operate more like a radio telescope. For example, the James Clerk Maxwell Telescope observes from wavelengths from 3 µm (0.003 mm) to 2000 µm (2 mm), but uses a parabolic aluminum antenna. On the other hand, the Spitzer Space Telescope, observing from about 3 µm (0.003 mm) to 180 µm (0.18 mm) uses a mirror (reflecting optics). Also using reflecting optics, the Hubble Space Telescope with Wide Field Camera 3 can observe in the frequency range from about 0.2 µm (0.0002 mm) to 1.7 µm (0.0017 mm) (from ultra-violet to infrared light).

At the photon energy of shorter wavelengths (higher frequency), glancing-incident optics, rather than fully reflecting optics are used. Telescopes such as TRACE and SOHO use special mirrors to reflect Extreme ultraviolet, producing higher resolution and brighter images than are otherwise possible. A larger aperture does not just mean that more light is collected, it also enables a finer angular resolution.

Telescopes may also be classified by location: ground telescope, space telescope, or flying telescope. They may also be classified by whether they are operated by professional astronomers or amateur astronomers. A vehicle or permanent campus containing one or more telescopes or other instruments is called an observatory.

Optical Telescopes

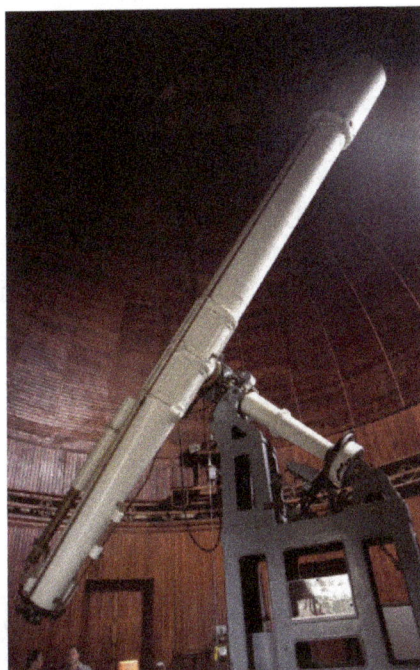

50 cm refracting telescope at Nice Observatory.

An optical telescope gathers and focuses light mainly from the visible part of the electromagnetic spectrum (although some work in the infrared and ultraviolet). Optical telescopes increase the apparent angular size of distant objects as well as their apparent brightness. In order for the image to be observed, photographed, studied, and sent to a computer, telescopes work by employing one or more curved optical elements, usually made from glass lenses and/or mirrors, to gather light and other electromagnetic radiation to bring that light or radiation to a focal point. Optical telescopes are used for astronomy and in many non-astronomical instruments, including: *theodolites* (including *transits*), *spotting scopes, monoculars, binoculars, camera lenses*, and *spyglasses*. There are three main optical types:

- The refracting telescope which uses lenses to form an image.

- The reflecting telescope which uses an arrangement of mirrors to form an image.

- The catadioptric telescope which uses mirrors combined with lenses to form an image.

Beyond these basic optical types there are many sub-types of varying optical design classified by the task they perform such as astrographs, comet seekers, solar telescope, etc.

Radio Telescopes

Radio telescopes are directional radio antennas used for radio astronomy. The dishes are sometimes constructed of a conductive wire mesh whose openings are smaller than the wavelength being observed. Multi-element Radio telescopes are constructed from pairs or larger groups of these dishes to synthesize large 'virtual' apertures that are similar in size to the separation between the telescopes; this process is known as aperture synthesis. As of 2005, the current record array size is many times the width of the Earth—utilizing space-based Very Long Baseline Interferometry (VLBI) telescopes such as the Japanese HALCA (Highly Advanced Laboratory for Communications and Astronomy) VSOP (VLBI Space Observatory Program) satellite. Aperture synthesis is now also being applied to optical telescopes using optical interferometers (arrays of optical telescopes) and aperture masking interferometry at single reflecting telescopes. Radio telescopes are also used to collect microwave radiation, which is used to collect radiation when any visible light is obstructed or faint, such as from quasars. Some radio telescopes are used by programs such as SETI and the Arecibo Observatory to search for extraterrestrial life.

X-ray Telescopes

Einstein Observatory was a space-based focusing optical X-ray telescope from 1978.

X-ray telescopes can use X-ray optics, such as a Wolter telescopes composed of ring-shaped 'glancing' mirrors made of heavy metals that are able to reflect the rays just a few degrees. The mirrors are usually a section of a rotated parabola and a hyperbola, or ellipse. In 1952, Hans Wolter outlined 3 ways a telescope could be built using only this kind of mirror. Examples of an observatory using this type of telescope are the Einstein Observatory, ROSAT, and the Chandra X-Ray Observatory. By 2010, Wolter focusing X-ray telescopes are possible up to 79 keV.

Gamma-ray Telescopes

Higher energy X-ray and Gamma-ray telescopes refrain from focusing completely and use coded aperture masks: the patterns of the shadow the mask creates can be reconstructed to form an image.

X-ray and Gamma-ray telescopes are usually on Earth-orbiting satellites or high-flying balloons since the Earth's atmosphere is opaque to this part of the electromagnetic spectrum. However, high energy X-rays and gamma-rays do not form an image in the same way as telescopes at visible wavelengths. An example of this type of telescope is the Fermi Gamma-ray Space Telescope.

The detection of very high energy gamma rays, with shorter wavelength and higher frequency than regular gamma rays, requires further specialization. An example of this type of observatory is VERITAS. Very high energy gamma-rays are still photons, like visible light, whereas cosmic rays includes particles like electrons, protons, and heavier nuclei.

A discovery in 2012 may allow focusing gamma-ray telescopes. At photon energies greater than 700 keV, the index of refraction starts to increase again.

High-energy Particle Telescopes

High-energy astronomy requires specialized telescopes to make observations since most of these particles go through most metals and glasses.

In other types of high energy particle telescopes there is no image-forming optical system. Cosmic-ray telescopes usually consist of an array of different detector types spread out over a large area. A Neutrino telescope consists of a large mass of water or ice, surrounded by an array of sensitive light detectors known as photomultiplier tubes. Originating direction of the neutrinos is determined by reconstructing the path of secondary particles scattered by neutrino impacts, from their interaction with multiple detectors. Energetic neutral atom observatories like Interstellar Boundary Explorer detect particles traveling at certain energies.

Other Types of Telescopes

Astronomy is not limited to using electromagnetic radiation. Additional information can be obtained using other media. The detectors used to observe the Universe are analogous to telescopes, these are:

- Gravitational-wave detector, the equivalent of a gravitational wave telescope, used for gravitational-wave astronomy.

- Neutrino detector, the equivalent of a neutrino telescope, used for neutrino astronomy.

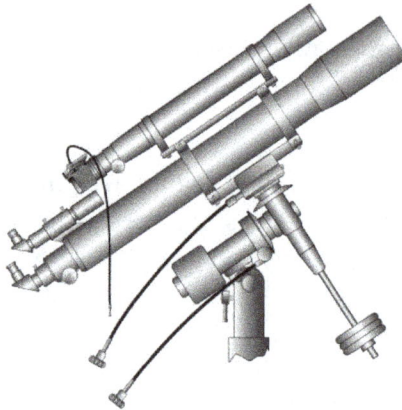

Equatorial-mounted Keplerian telescope

Types of Mount

A telescope mount is a mechanical structure which supports a telescope. Telescope mounts are designed to support the mass of the telescope and allow for accurate pointing of the instrument. Many sorts of mounts have been developed over the years, with the majority of effort being put into systems that can track the motion of the stars as the Earth rotates. The two main types of tracking mount are:

- Altazimuth mount

- Equatorial mount

Atmospheric Electromagnetic Opacity

Since the atmosphere is opaque for most of the electromagnetic spectrum, only a few bands can be observed from the Earth's surface. These bands are visible – near-infrared and a portion of the radio-wave part of the spectrum. For this reason there are no X-ray or far-infrared ground-based telescopes as these have to be observed from orbit. Even if a wavelength is observable from the ground, it might still be advantageous to place a telescope on a satellite due to astronomical seeing.

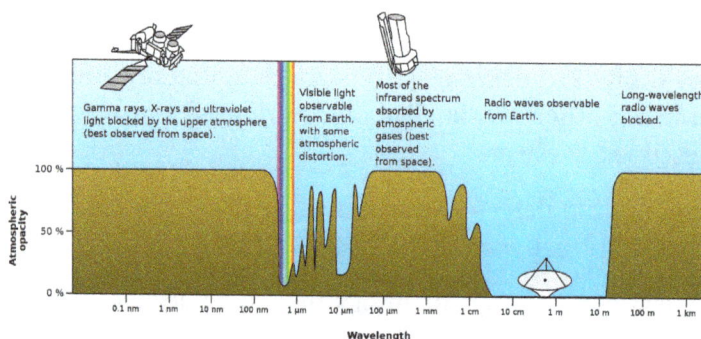

A diagram of the electromagnetic spectrum with the Earth's atmospheric transmittance (or opacity) and the types of telescopes used to image parts of the spectrum.

Telescopic Image from Different Telescope Types

Different types of telescope, operating in different wavelength bands, provide different information about the same object. Together they provide a more comprehensive understanding.

Crab Nebula: Remnant of an Exploded Star (Supernova)

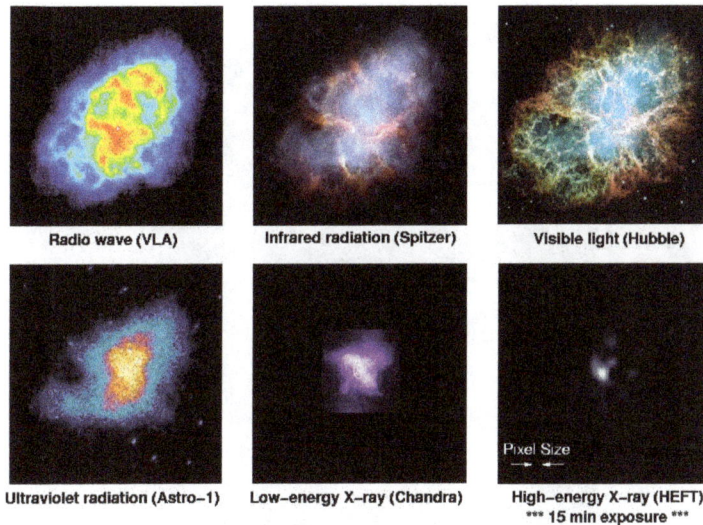

A 6′ wide view of the Crab nebula supernova remnant, viewed at different wavelengths of light by various telescopes

By Spectrum

Telescopes that operate in the electromagnetic spectrum:

Name	Telescope	Astronomy	Wavelength
Radio	Radio telescope	Radio astronomy (Radar astronomy)	more than 1 mm
Submillimetre	Submillimetre telescopes*	Submillimetre astronomy	0.1 mm – 1 mm
Far Infrared	–	Far-infrared astronomy	30 μm – 450 μm
Infrared	Infrared telescope	Infrared astronomy	700 nm – 1 mm
Visible	Visible spectrum telescopes	Visible-light astronomy	400 nm – 700 nm
Ultraviolet	Ultraviolet telescopes*	Ultraviolet astronomy	10 nm – 400 nm

X-ray	X-ray telescope	X-ray astronomy	0.01 nm – 10 nm
Gamma-ray	–	Gamma-ray astronomy	less than 0.01 nm

*Links to categories.

Optical Telescope

Eight-inch refracting telescope at Chabot Space and Science Center

An optical telescope is a telescope that gathers and focuses light, mainly from the visible part of the electromagnetic spectrum, to create a magnified image for direct view, or to make a photograph, or to collect data through electronic image sensors.

There are three primary types of optical telescope:

- refractors, which use lenses (dioptrics)

- reflectors, which use mirrors (catoptrics)

- catadioptric telescopes, which combine lenses and mirrors

A telescope's light gathering power and ability to resolve small detail is directly related to the diameter (or aperture) of its objective (the primary lens or mirror that collects and focuses the light). The larger the objective, the more light the telescope collects and the finer detail it resolves.

People use telescopes and binoculars for activities such as observational astronomy, ornithology, pilotage and reconnaissance, and watching sports or performance arts.

History

The telescope is more a discovery of optical craftsmen than an invention of a scientist. The lens and the properties of refracting and reflecting light had been known since antiquity and theory on how they worked were developed by ancient Greek philosophers, preserved and expanded on in the medieval Islamic world, and had reached a significantly advanced state by the time of the telescope's invention in early modern Europe. But the most significant step cited in the invention of the telescope was the development of lens manufacture for spectacles, first in Venice and Florence in the thirteenth century, and later in the spectacle making centers in both the Netherlands and Germany. It is in the Netherlands in 1608 where the first recorded optical telescopes (refracting telescopes) appeared. The invention is credited to the spectacle makers Hans Lippershey and Zacharias Janssen in Middelburg, and the instrument-maker and optician Jacob Metius of Alkmaar.

Galileo greatly improved on these designs the following year, and is generally credited as the first to use a telescope for astronomy. Galileo's telescope used Hans Lippershey's design of a convex objective lens and a concave eye lens, and this design is now called a Galilean telescope. Johannes Kepler proposed an improvement on the design that used a convex eyepiece, often called the Keplerian Telescope.

The next big step in the development of refractors was the advent of the Achromatic lens in the early 18th century, which corrected the chromatic aberration in Keplerian telescopes up to that time—allowing for much shorter instruments with much larger objectives.

For reflecting telescopes, which use a curved mirror in place of the objective lens, theory preceded practice. The theoretical basis for curved mirrors behaving similar to lenses was probably established by Alhazen, whose theories had been widely disseminated in Latin translations of his work. Soon after the invention of the refracting telescope Galileo, Giovanni Francesco Sagredo, and others, spurred on by their knowledge that curved mirrors had similar properties as lenses, discussed the idea of building a telescope using a mirror as the image forming objective. The potential advantages of using parabolic mirrors (primarily a reduction of spherical aberration with elimination of chromatic aberration) led to several proposed designs for reflecting telescopes, the most notable of which was published in 1663 by James Gregory and came to be called the Gregorian telescope, but no working models were built. Isaac Newton has been generally credited with constructing the first practical reflecting telescopes, the Newtonian telescope, in 1668 although due to their difficulty of construction and the poor performance of the speculum metal mirrors used it took over 100 years for reflectors to become popular. Many of the advances in reflecting telescopes included the perfection of parabolic mirror fabrication in the 18th century, silver coated glass mirrors in the 19th century, long-lasting aluminum coatings in the 20th century, segmented mirrors to allow larger diameters, and active optics to compensate for gravitational deformation. A mid-20th century innovation was catadioptric telescopes such as the Schmidt camera, which uses both a lens (corrector plate) and mirror as primary optical elements, mainly used for wide field imaging without spherical aberration.

The late 20th century has seen the development of adaptive optics and space telescopes to overcome the problems of astronomical seeing.

Principles

The basic scheme is that the primary light-gathering element the objective (1) (the convex lens or concave mirror used to gather the incoming light), focuses that light from the distant object (4) to a focal plane where it forms a real image (5). This image may be recorded or viewed through an eyepiece (2), which acts like a magnifying glass. The eye (3) then sees an inverted magnified virtual image (6) of the object.

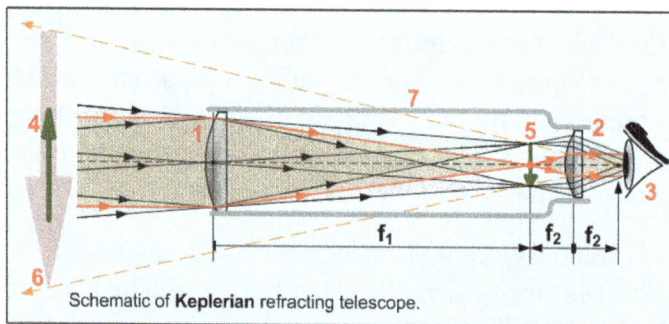

Schematic of **Keplerian** refracting telescope.

Schematic of a Keplerian refracting telescope. The arrow at (4) is a (notional) representation of the original image; the arrow at (5) is the inverted image at the focal plane; the arrow at (6) is the virtual image that forms in the viewer's visual sphere. The red rays produce the midpoint of the arrow; two other sets of rays (each black) produce its head and tail.

Inverted Images

Most telescope designs produce an inverted image at the focal plane; these are referred to as *inverting telescopes*. In fact, the image is both turned upside down and reversed left to right, so that altogether it is rotated by 180 degrees from the object orientation. In astronomical telescopes the rotated view is normally not corrected, since it does not affect how the telescope is used. However, a mirror diagonal is often used to place the eyepiece in a more convenient viewing location, and in that case the image is erect, but still reversed left to right. In terrestrial telescopes such as spotting scopes, monoculars and binoculars, prisms (e.g., Porro prisms) or a relay lens between objective and eyepiece are used to correct the image orientation. There are telescope designs that do not present an inverted image such as the Galilean refractor and the Gregorian reflector. These are referred to as *erecting telescopes*.

Design Variants

Many types of telescope fold or divert the optical path with secondary or tertiary mirrors. These may be integral part of the optical design (Newtonian telescope, Cassegrain reflector or similar types), or may simply be used to place the eyepiece or detector at a more convenient position. Telescope designs may also use specially designed additional lenses or mirrors to improve image quality over a larger field of view.

Characteristics

Design specifications relate to the characteristics of the telescope and how it performs optically. Several properties of the specifications may change with the equipment or accessories used with the telescope; such as Barlow lenses, star diagonals and eyepieces. These interchangeable accessories don't alter the specifications of the telescope, however they alter the way the telescopes properties function, typically magnification, angular resolution and FOV.

Surface Resolvability

The smallest resolvable surface area of an object, as seen through an optical telescope, is the limited physical area that can be resolved. It is analogous to angular resolution, but differs in definition: instead of separation ability between point-light sources it refers to the physical area that can be resolved. A familiar way to express the characteristic is the resolvable ability of features such as Moon craters or Sun spots. Expression using the formula is given by the sum of twice the resolving power R over aperture diameter D multiplied by the objects diameter D_{ob} multiplied by the constant Φ all divided by the objects apparent diameter D_a.

Resolving power R is derived from the wavelength λ using the same unit as aperture; where 550 nm to mm is given by: $R = \dfrac{\lambda}{10^6} = \dfrac{550}{10^6} = 0.00055..$

The constant Φ is derived from radians to the same unit as the objects apparent diameter; where the Moons apparent diameter of D_a $\dfrac{313\Pi}{10800}$ radians to arcsecs is given by:

$$D_a = \frac{313\Pi}{10800} * 206265 = 1878.$$

An example using a telescope with an aperture of 130 mm observing the Moon in a 550

nm wavelength, is given by: $F = \dfrac{\dfrac{2R}{D} * D_{ob} * \Phi}{D_a} = \dfrac{\dfrac{2*0.00055}{130} * 3474.2 * 206265}{1878} \approx 3.22$

The unit used in the object diameter results in the smallest resolvable features at that unit. In the above example they are approximated in kilometers resulting in the smallest resolvable Moon craters being 3.22 km in diameter. The Hubble Space Telescope has a primary mirror aperture of 2400 mm that provides a surface resolvability of Moon craters being 174.9 meters in diameter, or Sun spots of 7365.2 km in diameter.

Angular Resolution

Ignoring blurring of the image by turbulence in the atmosphere (atmospheric seeing) and optical imperfections of the telescope, the angular resolution of an optical telescope is determined by the diameter of the primary mirror or lens gathering the light (also termed its "aperture").

The Rayleigh criterion for the resolution limit α_R (in radians) is given by

$$\sin(\alpha_R) = 1.22\frac{\lambda}{D}$$

where λ is the wavelength and D is the aperture. For visible light (λ = 550 nm) in the small-angle approximation, this equation can be rewritten:

$$\alpha_R = \frac{138}{D}$$

Here, α_R denotes the resolution limit in arcseconds and D is in millimeters. In the ideal case, the two components of a double star system can be discerned even if separated by slightly less than α_R. This is taken into account by the Dawes limit

$$\alpha_D = \frac{116}{D}$$

The equation shows that, all else being equal, the larger the aperture, the better the angular resolution. The resolution is not given by the maximum magnification (or "power") of a telescope. Telescopes marketed by giving high values of the maximum power often deliver poor images.

For large ground-based telescopes, the resolution is limited by atmospheric seeing. This limit can be overcome by placing the telescopes above the atmosphere, e.g., on the summits of high mountains, on balloon and high-flying airplanes, or in space. Resolution limits can also be overcome by adaptive optics, speckle imaging or lucky imaging for ground-based telescopes.

Recently, it has become practical to perform aperture synthesis with arrays of optical telescopes. Very high resolution images can be obtained with groups of widely spaced smaller telescopes, linked together by carefully controlled optical paths, but these interferometers can only be used for imaging bright objects such as stars or measuring the bright cores of active galaxies.

Focal Length and Focal Ratio

The focal length of an optical system is a measure of how strongly the system converges or diverges light. For an optical system in air, it is the distance over which initially collimated rays are brought to a focus. A system with a shorter focal length has greater optical power than one with a long focal length; that is, it bends the rays more strongly, bringing them to a focus in a shorter distance. In astronomy, the f-number is commonly referred to as the *focal ratio* notated as N. The focal ratio of a telescope is defined as the focal length f of an objective divided by its diameter D or by the diameter of an aperture stop in the system. The focal length controls the field of view of the instrument and the scale of the image that is presented at the focal plane to an eyepiece, film plate, or CCD.

An example of a telescope with a focal length of 1200 mm and aperture diameter of 254 mm is given by: $N = \dfrac{f}{D} = \dfrac{1200}{254} \approx 4.7$

Numerically large Focal ratios are said to be *long* or *slow*. Small numbers are *short* or *fast*. There are no sharp lines for determining when to use these terms, an individual may consider their own standards of determination. Among contemporary astronomical telescopes, any telescope with a focal ratio slower (bigger number) than f/12 is generally considered slow, and any telescope with a focal ratio faster (smaller number) than f/6, is considered fast. Faster systems often have more optical aberrations away from the center of the field of view and are generally more demanding of eyepiece designs than slower ones. A fast system is often desired for practical purposes in astrophotography with the purpose of gathering more photons in a given time period than a slower system, allowing time lapsed photography to process the result faster.

Wide-field telescopes (such as astrographs), are used to track satellites and asteroids, for cosmic-ray research, and for astronomical surveys of the sky. It is more difficult to reduce optical aberrations in telescopes with low f-ratio than in telescopes with larger f-ratio.

Light-gathering Power

The light-gathering power of an optical telescope, also referred to as light grasp or aperture gain, is the ability of a telescope to collect a lot more light than the human eye. Its light-gathering power is probably its most important feature. The telescope acts as a *light bucket*, collecting all of the photons that come down on it from a far away object, where a larger bucket catches more photons resulting in more received light in a given time period, effectively brightening the image. This is why the pupils of your eyes enlarge at night so that more light reaches the retinas. The gathering power P compared against a human eye is the squared result of the division of the aperture D over the observer's pupil diameter D_p, with an average adult having a pupil diameter of 7mm. Younger persons host larger diameters, typically said to be 9mm, as the diameter of the pupil decreases with age.

An example gathering power of an aperture with 254 mm compared to an adult pupil diameter being 7 mm is given by:

$$P = \left(\frac{D}{D_p}\right)^2 = \left(\frac{254}{7}\right)^2 \approx 1316.7$$

Light-gathering power can be compared between telescopes by comparing the areas A of the two different apertures.

As an example, the light-gathering power of a 10 meter telescope is 25x that of a 2 meter telescope:

$$p = \frac{A_1}{A_2} = \frac{\pi 5^2}{\pi 1^2} = 25$$

For a survey of a given area, the field of view is just as important as raw light gathering power. Survey telescopes such as the Large Synoptic Survey Telescope try to maximize the product of mirror area and field of view (or etendue) rather than raw light gathering ability alone.

Magnification

The magnification through a telescope magnifies a viewing object while limiting the FOV. Magnification is often misleading as the optical power of the telescope, its characteristic is the most misunderstood term used to describe the observable world. At higher magnifications the image quality significantly reduces, usage of a Barlow lens—which increases the effective focal length of an optical system—multiplies image quality reduction.

Similar minor effects may be present when using star diagonals, as light travels through a multitude of lenses that increase or decrease effective focal length. The quality of the image generally depends on the quality of the optics (lenses) and viewing conditions— not on magnification.

Magnification itself is limited by optical characteristics. With any telescope or micro-scope, beyond a practical maximum magnification, the image looks bigger but shows no more detail. It occurs when the finest detail the instrument can resolve is magnified to match the finest detail the eye can see. Magnification beyond this maximum is some-times called *empty magnification*.

To get the most detail out of a telescope, it is critical to choose the right magnification for the object being observed. Some objects appear best at low power, some at high power, and many at a moderate magnification. There are two values for magnification, a minimum and maximum. A wider field of view eyepiece may be used to keep the same eyepiece focal length whilst providing the same magnification through the telescope. For a good quality telescope operating in good atmospheric conditions, the maximum usable magnification is limited by diffraction.

Visual

The visual magnification M of the field of view through a telescope can be determined by the telescopes focal length f divided by the eyepiece focal length f_e (or diameter). The maximum is limited by the diameter of the eyepiece.

An example of visual magnification using a telescope with a 1200 mm focal length and 3 mm eyepiece is given by:

$$M = \frac{f}{f_e} = \frac{1200}{3} = 400$$

Minimum

There is a lowest usable magnification on a telescope. The increase in brightness with reduced magnification has a limit related to something called the exit pupil. The exit pupil is the cylinder of light coming out of the eyepiece, hence the lower the magnification, the larger the exit pupil. The minimum M_m an be calculated by dividing the telescope aperture D over the exit pupil diameter D_{ep}. Decreasing the magnification past this limit cannot increase brightness, at this limit there is no benefit for decreased magnification. Likewise calculating the exit pupil D_{ep} is a division of the aperture diameter D and the visual magnification M used. The minimum often may not be reachable with some telescopes, a telescope with a very long focal length may require a *longer-focal-length* eyepiece than is possible.

An example of the lowest usable magnification using a 254 mm aperture and 7 mm exit pupil is given by: $M_m = \frac{D}{D_{ep}} = \frac{254}{7} \approx 36,$ whilst the exit pupil diameter using a 254 mm aperture and 36x magnification is given by:

$$D_{ep} = \frac{D}{M} = \frac{254}{36} \approx 7$$

Optimum

A useful reference is:

- For small objects with low surface brightness (such as galaxies), use a moderate magnification.

- For small objects with high surface brightness (such as planetary nebulae), use a high magnification.

- For large objects regardless of surface brightness (such as diffuse nebulae), use low magnification, often in the range of minimum magnification.

Only personal experience determines the best optimum magnifications for objects, relying on observational skills and seeing conditions.

Field of View

Field of view is the extent of the observable world seen at any given moment, through an instrument (e.g., telescope or binoculars), or by naked eye. There are various expressions of field of view, being a specification of an eyepiece or a characteristic determined from and eyepiece and telescope combination. A physical limit derives from the combination where the FOV cannot be viewed larger than a defined maximum, due to diffraction of the optics.

Apparent

Apparent FOV is the observable world observed through an ocular eyepiece without insertion into a telescope. It is limited by the barrel size used in a telescope, generally with modern telescopes that being either 1.25 or 2 inches in diameter. A wider FOV may be used to achieve a more vast observable world given the same magnification compared with a smaller FOV without compromise to magnification. Note that increasing the FOV lowers surface brightness of an observed object, as the gathered light is spread over more area, in relative terms increasing the observing area proportionally lowers surface brightness dimming the observed object. Wide FOV eyepieces work best at low magnifications with large apertures, where the relative size of an object is viewed at higher comparative standards with minimal magnification giving an overall brighter image to begin with.

True

True FOV is the observable world observed though an ocular eyepiece inserted into a telescope. Knowing the true FOV of eyepieces is very useful since it can be used to compare what is seen through the eyepiece to printed or computerized star charts that help identify what is observed. True FOV V_t is the division of apparent FOV V_a over magnification M.

An example of true FOV using an eyepiece with 52° apparent FOV used at 81.25x magnification is given by:

$$V_t = \frac{V_a}{M} = \frac{52}{81.25} = 0.64°$$

Maximum

Max FOV is a term used to describe the maximum useful true FOV limited by the optics of the telescope, it is a physical limitation where increases beyond the maximum remain at maximum. Max FOV V_m is the barrel size B over the telescopes focal length f converted from radian to degrees.

An example of max FOV using a telescope with a barrel size of 31.75 mm (1.25 inches) and focal length of 1200 mm is given by:

$$v_m = B * \frac{\frac{180}{\pi}}{f} \approx 31.75 * \frac{57.2958}{1200} \approx 1.52°$$

Observing Through a Telescope

There are many properties of optical telescopes and the complexity of observation using one can be a daunting task, experience and experimentation are the major contributors to understanding how to maximize your observations. In practice, only two main properties of a telescope determine how observation differs: the focal length and aperture. These relate as to how the optical system views an object or range and how much light is gathered through an ocular eyepiece. Eyepieces further determine how the field of view and magnification of the observable world change.

Observable World

This term describes what can be seen using a telescope, when viewing an object or range the observer may use many different techniques. Understanding what can be viewed and how to view it depends on the field of view. Viewing an object at a size that fits entirely in the field of view is measured using the two telescope properties—focal length and aperture, with the inclusion of an ocular eyepiece with suitable focal length (or diameter). Comparing the observable world and the angular diameter of an object shows how much of the object we see. However, the relationship with the optical system may not result in high surface brightness. Celestial objects are often dim because of their vast distance, and detail may be limited by diffraction or unsuitable optical properties.

Field of View and Magnification Relationship

Finding what can be seen through the optical system begins with the eyepiece providing our field of view and magnification, the magnification is given by the division of the telescope and eyepiece focal lengths. Using an example of an amateur telescope such as a Newtonian telescope with an aperture D of 130 mm (5") and focal length f of 650mm (25.5"), we use an eyepiece with a focal length d of 8 mm and apparent field of view v_a of 52°. The magnification at which the observable world is viewed at is given by: $M = \frac{f}{d} = \frac{650}{8} = 81.25.$ The true field of view v_t requires the magnification, which is formulated by its division over the apparent field of view: $v_t = \frac{v_a}{M} = \frac{52}{81.25} = 0.64.$. Our resulting true field of view is 0.64°, allowing an object such as the Orion nebula, which appears elliptical with an angular diameter of 65 x 60 arcminutes to be viewable through the telescope in its entirety, where the whole of the nebula is within the observable world. Using methods such as this, can greatly increase your viewing potential ensuring the observable world can contain the entire object, or whether to increase/decrease magnification viewing the object in a different aspect.

Brightness Factor

Note that the surface brightness at such a magnification significantly reduces, resulting in a far dimmer appearance. A dimmer appearance results in less visual detail of the object. Details such as matter, rings, spiral arms, and gases may be completely hidden from the observer, giving a far less *complete* view of the object or range. Physics dictates that at the theoretical minimum magnification of the telescope, the surface brightness is at 100%. Practically, however, various factors prevent 100% brightness. These include telescope limitations (focal length, eyepiece focal length,etc.) and the age of the observer.

Age plays a role in brightness, as a contributing factor is the observer's pupil. With age the pupil naturally shrinks in diameter, generally accepted a young adult may have a 7 mm diameter pupil, an older adult as little as 5 mm and a younger person larger at 9 mm. The minimum magnification m can be expressed as the division of the aperture D and pupil P diameter given by: $m = \dfrac{D}{d} = \dfrac{130}{7} \approx 18.6$. A problematic instance may be apparent achieving a theoretical surface brightness of 100%, as the required effective focal length of the optical system may require an eyepiece with too large a diameter.

Some telescopes cannot achieve the theoretical surface brightness of 100%, while some telescopes can achieve it using a very small diameter eyepiece. To find what eyepiece is required to get our minimum magnification we can rearrange the magnification formula, where its now the division of the telescopes focal length over the minimum magnification: $\dfrac{F}{m} = \dfrac{650}{18.6} \approx 35$. An eyepiece of 35 mm is a non-standard size and would not be purchasable, in this scenario to achieve 100% we would required a standard manufactured eyepiece size of 40 mm. As the eyepiece has a larger focal length than our minimum magnification, an abundance of wasted light isn't received through our eyes.

Exit Pupil

The increase in surface brightness as you reduce magnification is limited, that limitation is what we describe as the exit pupil; a cylinder of light that projects out the eyepiece to the observer. An exit pupil must match or be smaller in diameter than our pupil to receive the full amount of projected light, a larger exit pupil results in the wasted light. The exit pupil e can be derived with from division of the telescope aperture D and the minimum magnification m, derived by: $me = \dfrac{D}{m} = \dfrac{130}{18.6} \approx 7.$. The pupil and exit pupil are almost identical in diameter giving no wasted observable light with the optical system. A 7mm pupil falls slightly short of 100% brightness, where the surface brightness B can be measured from the product of the constant 2, by the square of the pupil p resulting in: $B = 2 * p^2 = 2 * 7^2 = 98.$. The limitation here is the pupil diameter, it's an unfortunate result and degrades with age. Some observable light loss is expected and decreasing the magnification cannot not increase surface brightness once the system has reached its minimum usable magnification, hence why the term is referred to as *usable*.

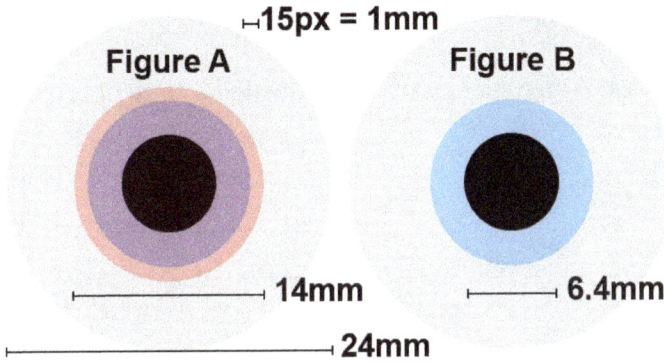

15px = 1mm

Figure A **Figure B**

14mm 6.4mm

24mm

These eyes represent a scaled figure of the human eye where 15 px = 1 mm, they have a pupil diameter of 7 mm. *Figure A* has an exit pupil diameter of 14 mm, which for astronomy purposes results in a 75% loss of light. *Figure B* has an exit pupil of 6.4 mm, which allows the full 100% of observable light to be perceived by the observer.

Image Scale

When using a CCD to record observations, the CCD is placed in the focal plane. Image scale (sometimes called *plate scale*) describes how the angular size of the object being observed is related to the physical size of the projected image in the focal plane

$$i = \frac{\alpha}{s},$$

where i is the image scale, α is the angular size of the observed object, and s is the physical size of the projected image. In terms of focal length image scale is

$$i = \frac{1}{f},$$

where i is measured in radians per meter (rad/m), and f is measured in meters. Normally i is given in units of arcseconds per millimeter ("/mm). So if the focal length is measured in millimeters, the image scale is

$$i("/mm) = \frac{1}{f(mm)} \left[\frac{180 \times 3600}{\pi} \right].$$

The derivation of this equation is fairly straightforward and the result is the same for reflecting or refracting telescopes. However, conceptually it is easier to derive by considering a reflecting telescope. If an extended object with angular size α is observed through a telescope, then due to the Laws of reflection and Trigonometry the size of the image projected onto the focal plane will be

$$s = \tan(\alpha)f.$$

Thefore, the image scale (angular size of object divided by size of projected image) will be

$$i = \frac{\alpha}{s} = \frac{\alpha}{\tan(\alpha)f},$$

and by using the small angle relation $\tan(a) \approx a$, when $a \ll 1$ (N.B. only valid if a is in radians), we obtain

$$i = \frac{\alpha}{\alpha f} = \frac{1}{f}.$$

Imperfect Images

No telescope can form a perfect image. Even if a reflecting telescope could have a perfect mirror, or a refracting telescope could have a perfect lens, the effects of aperture diffraction are unavoidable. In reality, perfect mirrors and perfect lenses do not exist, so image aberrations in addition to aperture diffraction must be taken into account. Image aberrations can be broken down into two main classes, monochromatic, and polychromatic. In 1857, Philipp Ludwig von Seidel (1821–1896) decomposed the first order monochromatic aberrations into five constituent aberrations. They are now commonly referred to as the five Seidel Aberrations.

The Five Seidel Aberrations

Spherical aberration

> The difference in focal length between paraxial rays and marginal rays, proportional to the square of the objective diameter.

Coma

> A defect by which points appear as comet-like asymmetrical patches of light with tails, which makes measurement very imprecise. Its magnitude is usually deduced from the optical sine theorem.

Astigmatism

> The image of a point forms focal lines at the sagittal and tangential foci and in between (in the absence of coma) an elliptical shape.

Curvature of Field

> The Petzval field curvature means that the image, instead of lying in a plane, actually lies on a curved surface, described as hollow or round. This causes problems

when a flat imaging device is used e.g., a photographic plate or CCD image sensor.

Distortion

> Either barrel or pincushion, a radial distortion that must be corrected when combining multiple images (similar to stitching multiple photos into a panoramic photo).

Optical defects are always listed in the above order, since this expresses their interdependence as first order aberrations via moves of the exit/entrance pupils. The first Seidel aberration, Spherical Aberration, is independent of the position of the exit pupil (as it is the same for axial and extra-axial pencils). The second, coma, changes as a function of pupil distance and spherical aberration, hence the well-known result that it is impossible to correct the coma in a lens free of spherical aberration by simply moving the pupil. Similar dependencies affect the remaining aberrations in the list.

Chromatic Aberrations

> Longitudinal chromatic aberration: As with spherical aberration this is the same for axial and oblique pencils.

> Transverse chromatic aberration (chromatic aberration of magnification)

Astronomical Research Telescopes

Two of the four Unit Telescopes that make up the ESO's VLT, on a remote mountaintop, 2600 metres above sea level in the Chilean Atacama Desert.

Optical telescopes have been used in astronomical research since the time of their invention in the early 17th century. Many types have be constructed over the years depending on the optical technology, such as refracting and reflecting, the nature of the light or object being imaged, and even where they are placed, such as space telescopes. Some are classified by the task they perform such as Solar telescopes.

Large Reflectors

Nearly all large research-grade astronomical telescopes are reflectors. Some reasons are:

- In a lens the entire volume of material has to be free of imperfection and inhomogeneities, whereas in a mirror, only one surface has to be perfectly polished.

- Light of different colors travels through a medium other than vacuum at different speeds. This causes chromatic aberration.

- Reflectors work in a wider spectrum of light since certain wavelengths are absorbed when passing through glass elements like those found in a refractor or catadioptric.

- There are technical difficulties involved in manufacturing and manipulating large-diameter lenses. One of them is that all real materials sag in gravity. A lens can only be held by its perimeter. A mirror, on the other hand, can be supported by the whole side opposite to its reflecting face.

Most large research reflectors operate at different focal planes, depending on the type and size of the instrument being used. These including the prime focus of the main mirror, the cassegrain focus (light bounced back down behind the primary mirror), and even external to the telescope all together (such as the Nasmyth and coudé focus).

A new era of telescope making was inaugurated by the Multiple Mirror Telescope (MMT), with a mirror composed of six segments synthesizing a mirror of 4.5 meters diameter. This has now been replaced by a single 6.5 m mirror. Its example was followed by the Keck telescopes with 10 m segmented mirrors.

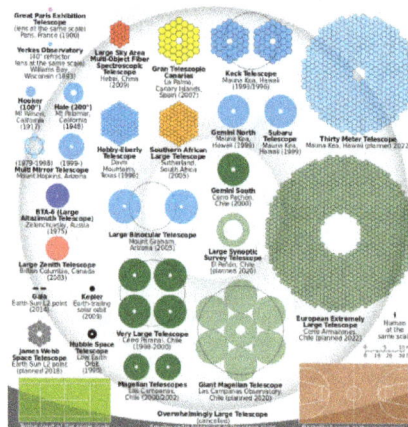

Comparison of nominal sizes of primary mirrors of some notable optical telescopes

The largest current ground-based telescopes have a primary mirror of between 6 and 11 meters in diameter. In this generation of telescopes, the mirror is usually very thin, and is kept in an optimal shape by an array of actuators. This technology has driven new designs for future telescopes with diameters of 30, 50 and even 100 meters.

Harlan J. Smith Telescope reflecting telescope at McDonald Observatory, Texas

Relatively cheap, mass-produced ~2 meter telescopes have recently been developed and have made a significant impact on astronomy research. These allow many astronomical targets to be monitored continuously, and for large areas of sky to be surveyed. Many are robotic telescopes, computer controlled over the internet, allowing automated follow-up of astronomical events.

Initially the detector used in telescopes was the human eye. Later, the sensitized photographic plate took its place, and the spectrograph was introduced, allowing the gathering of spectral information. After the photographic plate, successive generations of electronic detectors, such as the charge-coupled device (CCDs), have been perfected, each with more sensitivity and resolution, and often with a wider wavelength coverage.

Current research telescopes have several instruments to choose from such as:

- imagers, of different spectral responses

- spectrographs, useful in different regions of the spectrum

- polarimeters, that detect light polarization.

The phenomenon of optical diffraction sets a limit to the resolution and image quality that a telescope can achieve, which is the effective area of the Airy disc, which limits how close two such discs can be placed. This absolute limit is called the diffraction limit (and may be approximated by the Rayleigh criterion, Dawes limit or Sparrow's resolution limit). This limit depends on the wavelength of the studied light (so that the limit for red light comes much earlier than the limit for blue light) and on the diameter of the telescope mirror. This means that a telescope with a certain mirror diameter can theoretically resolve up to a certain limit at a certain wavelength. For conventional telescopes on Earth, the diffraction limit is not relevant for telescopes bigger than about 10 cm. Instead, the seeing, or blur caused by the atmosphere, sets the resolution limit. But in space, or if adaptive optics are used, then reaching the diffraction limit is sometimes possible. At this point, if greater resolution is needed at that wavelength, a wider mirror has to be built or aperture synthesis performed using an array of nearby telescopes.

In recent years, a number of technologies to overcome the distortions caused by atmosphere on ground-based telescopes have been developed, with good results.

Photometer

A photometer, generally, is an instrument that measures light intensity or optical properties of solutions or surfaces. Photometers measure:

- Illuminance

- Irradiance

- Light absorption

- Scattering of light

- Reflection of light

- Fluorescence

- Phosphorescence

- Luminescence

A photometer

History

Before electronic light sensitive elements were developed, photometry was done by estimation by the eye. The relative luminous flux of a source was compared with a standard source. The photometer is placed such that the illuminance from the source being investigated is equal to the standard source, as the human eye can judge equal illuminance. The relative luminous fluxes can then be calculated as the illuminance decreases proportionally to the inverse square of distance. A standard example of such a photometer consists of a piece of paper with an oil spot on it that makes the paper slightly more transparent. When the spot is not visible from either side, the illuminance from the two sides is equal.

By 1861, three types were in common use. These were Rumford's photometer, Ritchie's photometer, and photometers that used the extinction of shadows, which was considered to be the most precise.

Principle of Photometers

Most photometers detect the light with photoresistors, photodiodes or photomultipliers. To analyze the light, the photometer may measure the light after it has passed through a filter or through a monochromator for determination at defined wavelengths or for analysis of the spectral distribution of the light.

Photon Counting

Some photometers measure light by counting individual photons rather than incoming flux. The operating principles are the same but the results are given in units such as photons/cm² or photons·cm^{-2}·sr^{-1} rather than W/cm² or W·cm^{-2}·sr^{-1}.

Due to their individual photon counting nature, these instruments are limited to observations where the irradiance is low. The irradiance is limited by the time resolution of its associated detector readout electronics. With current technology this is in the megahertz range. The maximum irradiance is also limited by the throughput and gain parameters of the detector itself.

The light sensing element in photon counting devices in NIR, visible and ultraviolet wavelengths is a photomultiplier to achieve sufficient sensitivity.

In airborne and space-based remote sensing such photon counters are used at the upper reaches of the electromagnetic spectrum such as the X-ray to far ultraviolet. This is usually due to the lower radiant intensity of the objects being measured as well as the difficulty of measuring light at higher energies using its particle-like nature as compared to the wavelike nature of light at lower frequencies. Conversely, radiometers are typically used for remote sensing from the visible, infrared though radio frequency range.

Photography

Photometers are used to determine the correct exposure in photography. In modern cameras, the photometer is usually built in. As the illumination of different parts of the picture varies, advanced photometers measure the light intensity in different parts of the potential picture and use an algorithm to determine the most suitable exposure for the final picture, adapting the algorithm to the type of picture intended. Historically, a photometer was separate from the camera and known as an exposure meter. The advanced photometers then could be used either to measure the light from the potential picture as a whole, to measure from elements of the picture to ascertain that the most important parts of the picture are optimally exposed, or to measure the incident light to the scene with an integrating adapter.

Visible Light Reflectance Photometry

A reflectance photometer measures the reflectance of a surface as a function of wavelength. The surface is illuminated with white light, and the reflected light is measured after passing through a monochromator. This type of measurement has mainly practical applications, for instance in the paint industry to characterize the colour of a surface objectively.

UV and Visible Light Transmission Photometry

These are optical instruments for measurement of the absorption of light of a given wavelength (or a given range of wavelengths) of coloured substances in solution. From the light absorption, Beer's law makes it possible to calculate the concentration of the coloured substance in the solution. Due to its wide range of application and its reliability and robustness, the photometer has become one of the principal instruments in biochemistry and analytical chemistry. Absorption photometers for work in aqueous solution work in the ultraviolet and visible ranges, from wavelength around 240 nm up to 750 nm.

The principle of spectrophotometers and filter photometers is that (as far as possible) monochromatic light is allowed to pass through a container (cell) with optically flat windows containing the solution. It then reaches a light detector, that measures the intensity of the light compared to the intensity after passing through an identical cell with the same solvent but without the coloured substance. From the ratio between the light intensities, knowing the capacity of the coloured substance to absorb light (the absorbency of the coloured substance, or the photon cross section area of the molecules of the coloured substance at a given wavelength), it is possible to calculate the concentration of the substance using Beer's law.

Two types of photometers are used: spectrophotometer and filter photometer. In spectrophotometers a monochromator (with prism or with grating) is used to obtain monochromatic light of one defined wavelength. In filter photometers, optical filters are used to give the monochromatic light. Spectrophotometers can thus easily be set to measure the absorbance at different wavelengths, and they can also be used to scan the spectrum of the absorbing substance. They are in this way more flexible than filter photometers, also give a higher optical purity of the analyzing light, and therefore they are preferably used for research purposes. Filter photometers are cheaper, robuster and easier to use and therefore they are used for routine analysis. Photometers for microtiter plates are filter photometers.

Infrared Light Transmission Photometry

Spectrophotometry in infrared light is mainly used to study structure of substances, as given groups give absorption at defined wavelengths. Measurement in aqueous solution is generally not possible, as water absorbs infrared light strongly in some wavelength ranges. Therefore, infrared spectroscopy is either performed in the gaseous phase (for volatile substances) or with the substances pressed into tablets together with salts that are transparent in the infrared range. Potassium bromide (KBr) is commonly used for this purpose. The substance being tested is thoroughly mixed with specially purified KBr and pressed into a transparent tablet, that is placed in the beam of light. The analysis of the wavelength dependence is generally not done using a monochromator as it is in UV-Vis, but with the use of an interferometer. The interference pattern can be analyzed using a Fourier transform algorithm. In this way, the whole wavelength range can be analyzed simultaneously, saving time, and an interferometer is also less expensive than a monochromator. The light absorbed in the infrared region does not correspond to electronic excitation of the substance studied, but rather to different kinds of vibrational excitation. The vibrational excitations are characteristic of different groups in a molecule, that can in this way be identified. The infrared spectrum typically has very narrow absorption lines, which makes them unsuited for quantitative analysis but gives very detailed information about the molecules. The frequencies of the different modes of vibration varies with isotope, and therefore different isotopes give different peaks. This makes it possible also to study the isotopic composition of a sample with infrared spectrophotometry.

Atomic Absorption Photometry

Atomic absorption photometers are photometers that measure the light from a very hot flame. The solution to be analyzed is injected into the flame at a constant, known rate. Metals in the solution are present in atomic form in the flame. The monochromatic light in this type of photometer is generated by a discharge lamp where the discharge takes place in a gas with the metal to be determined. The discharge then emits light with wavelengths corresponding to the spectral lines of the metal. A filter may be used to isolate one of the main spectral lines of the metal to be analyzed. The light is absorbed by the metal in the flame, and the absorption is used to determine the concentration of the metal in the original solution.

Rumford's Photometer

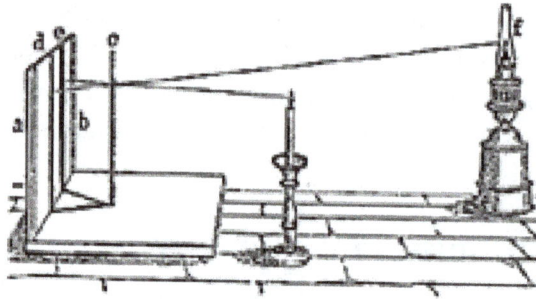

Rumford's Photometer

Rumford's photometer (also called a shadow photometer) depended on the principle that a brighter light would cast a deeper shadow. The two lights to be compared were used to cast a shadow onto paper. If the shadows were of the same depth, the difference in distance of the lights would indicate the difference in intensity (e.g. a light twice as far would be four times the intensity).

Ritchie's Photometer

Ritchie's photometer depended on equal illumination of surfaces. It consisted of a box (a,b) six or eight inches long, and one in width and depth. In the middle, a wedge of wood (f,e,g) was angled upwards and covered with white paper. The user's eye looked through a tube (d) at the top of a box. The height of the apparatus was also adjustable via the stand (c). The lights to compare were placed at the side of the box (m, n)—which illuminated the paper surfaces so that the eye saw both surfaces at once. By changing the position of the lights, they were made to illuminate both surfaces equally, with the difference in intensity corresponding to the square of the difference in distance.

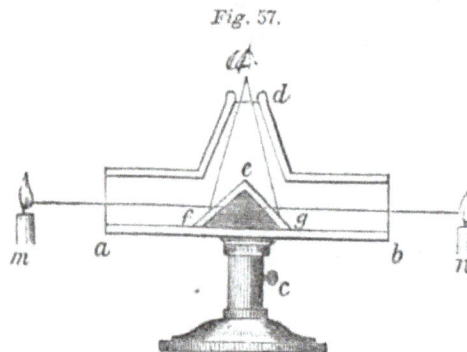

Fig. 57.

Ritchie's Photometer

Method of Extinction of Shadows

This type of photometer depended on the fact that if a light throws the shadow of an opaque object onto a white screen, there is a certain distance that, if a second light is brought there, obliterates all traces of the shadow.

Refractometer

A refractometer is a laboratory or field device for the measurement of an index of refraction (refractometry). The index of refraction is calculated from Snell's law while for mixtures, the index of refraction can be calculated from the composition of the material using several mixing rules such as the Gladstone–Dale relation and Lorentz–Lorenz equation.

Types of Refractometers

There are four main types of refractometers: traditional handheld refractometers, digital handheld refractometers, laboratory or Abbe refractometers, and inline process refractometers. There is also the Rayleigh Refractometer used (typically) for measuring the refractive indices of gases.

In veterinary medicine, a refractometer is used to measure the total plasma protein in a blood samples.

In drug diagnostics, a refractometer is used to measure the specific gravity of human urine.

In gemology/gemmology, the gemstone refractometer is one the fundamental pieces of equipment used in a gemological laboratory. Gemstones are transparent minerals and can therefore be examined using optical methods. Refractive index is a material constant, dependent on the chemical composition of a substance. The refractometer is used to help identify gem materials by measuring their refractive index, one of the principal properties used in determining the type of a gemstone. Due to the dependence of the refractive index on the wavelength of the light used (*i.e.* dispersion), the measurement is normally taken at the wavelength of the sodium line D-line (Na_D) of ~589 nm. This is either filtered out from daylight or generated with a monochromatic light-emitting diode (LED). Certain stones such as rubies, sapphires, tourmalines and topaz are optically anisotropic. They demonstrate birefringence based on the polarisation plane of the light. The two different refractive indexes are classified using a polarisation filter. Gemstone refractometers are available both as classic optical instruments and as electronic measurement devices with a digital display.

In marine aquarium keeping, a refractometer is used to measure the salinity and specific gravity of the water. In Automobile Industry keeping, a refractometer is used to measure the Coolant Concentration and Ph Value of the Coolant Oils for CNC Machining Process.

In homebrewing, a refractometer is used to measure the specific gravity before fermentation to determine the amount of fermentable sugars which will potentially be converted to alcohol.

In beekeeping, a refractometer is used to measure the amount of water in honey.

Gemology refractometer ER604 used to test light bending in gemstones; courtesy of
A.KRÜSS Optronic GmbH

A wine grape grower with refractometer

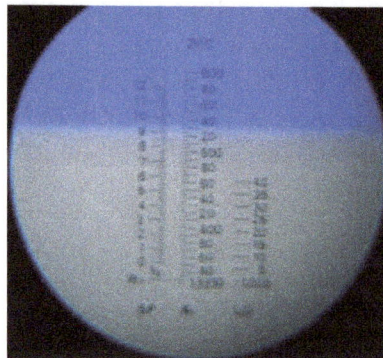

Density evaluation of abdominal fluid of a cat with feline infectious peritonitis
by a refractometer.

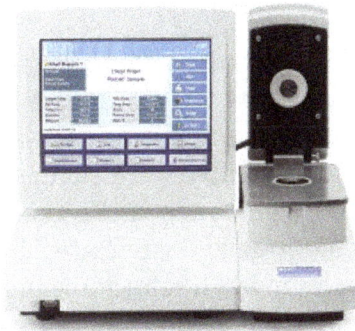

Automatic benchtop refractometer

Automatic Refractometers

Automatic refractometers automatically measure the refractive index of a sample. The automatic measurement of the refractive index of the sample is based on the determination of the critical angle of total reflection. A light source, usually a long-life LED, is focused onto a prism surface via a lens system. An interference filter guarantees the specified wavelength. Due to focusing light to a spot at the prism surface, a wide range of different angles is covered. As shown in the figure *"Schematic setup of an automatic refractometer"* the measured sample is in direct contact with the measuring prism. Depending on its refractive index, the incoming light below the critical angle of total reflection is partly transmitted into the sample, whereas for higher angles of incidence the light is totally reflected. This dependence of the reflected light intensity from the incident angle is measured with a high-resolution sensor array. From the video signal taken with the CCD sensor the refractive index of the sample can be calculated. This method of detecting the angle of total reflection is independent on the sample properties. It is even possible to measure the refractive index of optically dense strongly absorbing samples or samples containing air bubbles or solid particles . Furthermore, only a few microliters are required and the sample can be recovered. This determination of the refraction angle is independent of vibrations and other environmental disturbances.

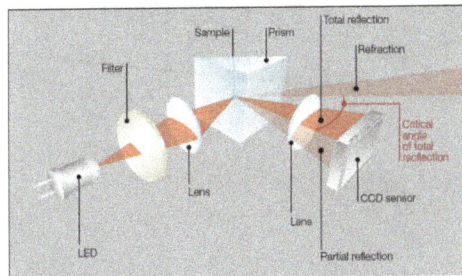

'Schematic setup of an automatic refractometer: An LED light source is imaged under a wide range of angles onto a prism surface which is in contact with a sample. Depending on the difference in the refractive index between prism material and sample the light is partly transmitted or totally reflected. The critical angle of total reflection is determined by measuring the reflected light intensity as a function of the incident angle - Source of image: Anton Paar GmbH, www.anton-paar.com

Influence of Wavelength

The refractive index of a given sample varies with wavelength for all materials. This dispersion relation is nonlinear and is characteristic for every material. In the visible range, a decrease of the refractive index comes with increasing wavelength. In glass prisms very little absorption is observable. In the infrared wavelength range several absorption maxima and fluctuations in the refractive index appear. To guarantee a high quality measurement with an accuracy of up to 0.00002 in the refractive index the wavelength has to be determined correctly. Therefore, in modern refractometers the wavelength is tuned to a bandwidth of +/-0.2 nm to ensure correct results for samples with different dispersions.

Modern Automatic Refractometers - Source of image: Anton Paar GmbH, www.anton-paar.com

Influence of Temperature

Refractometer with funnel type sampler - Flow cell with filling funnel - Source of image: Rudolph Research Analytical - www.rudolphresearch.com

Temperature has a very important influence on the refractive index measurement. Therefore, the temperature of the prism and the temperature of the sample have to be controlled with high precision. There are several subtly-different designs for controlling the temperature; but there are some key factors common to all, such as high-precision temperature sensors and Peltier devices to control the temperature of the sample and the prism. The temperature control accuracy of these devices should be designed so that the variation in sample temperature is small enough that it will not cause a detectable refractive-index change.

External water baths were used in the past but are no longer needed.

Extended Possibilities of Automatic Refractometers

Automatic refractometers are microprocessor-controlled electronic devices. This means they can have a high degree of automation and also be combined with other measuring devices

Flow Cells

There are different types of sample cells available, ranging from a flow cell for a few microliters to sample cells with a filling funnel for fast sample exchange without cleaning the measuring prism in between. The sample cells can also be used for the measurement of poisonous and toxic samples with minimum exposure to the sample. Micro cells require only a few microliters volume, assure good recovery of expensive samples and prevent evaporation of volatile samples or solvents. They can also be used in automated systems for automatic filling of the sample onto the refractometer prism. For convenient filling of the sample through a funnel, flow cells with a filling funnel are available. These are used for fast sample exchange in quality control applications.

Automatic Sample Feeding

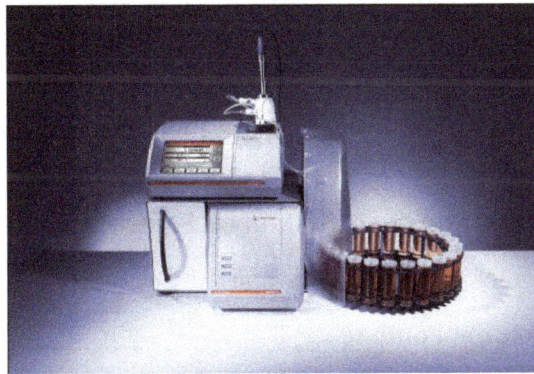

Automatic refractometer with sample changer for automatic measurement of a large number of samples - Source of image: Anton Paar GmbH, www.anton-paar.com

Once an automatic refractometer is equipped with a flow cell, the sample can either be filled by means of a syringe or by using a peristaltic pump. Modern refractometers have the option of a built-in peristaltic pump. This is controlled via the instrument's software menu. A peristaltic pump opens the way to monitor batch processes in the laboratory or perform multiple measurements on one sample without any user interaction. This eliminates human error and assures a high sample throughput.

If an automated measurement of a large number of samples is required, modern automatic refractometers can be combined with an automatic sample changer. The sample changer is controlled by the refractometer and assures fully automated measurements of the samples placed in the vials of the sample changer for measurements.

Multiparameter Measurements

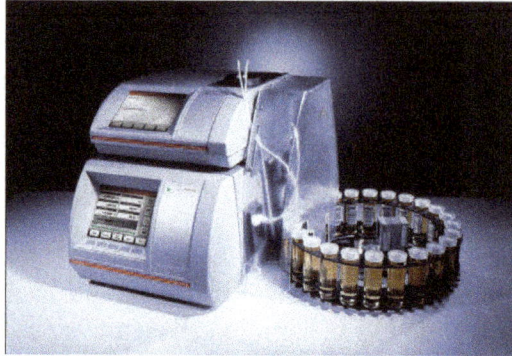

Measuring combination of an automatic refractometer and a density meter as widely used in the flavors and fragrances industry - Source of image: Anton Paar GmbH, www.anton-paar.com

Today's laboratories do not only want to measure the refractive index of samples, but several additional parameters like density or viscosity to perform efficient quality control. Due to the microprocessor control and a number of interfaces, automatic refractometers are able to communicate with computers or other measuring devices, e.g. density meters, pH meters or viscosity meters, to store refractive index data and density data (and other parameters) into one database.

Software Features

Automatic refractometers do not only measure the refractive index, but offer a lot of additional software features, like

- Instrument settings and configuration via software menu
- Automatic data recording into a database
- User-configurable data output
- Export of measuring data into Microsoft Excel data sheets
- Statistical functions
- Predefined methods for different kinds of applications
- Automatic checks and adjustments
- Check if sufficient amount of sample is on the prism
- Data recording only if the results are plausible

Pharma Documentation and Validation

Refractometers are often used in pharmaceutical applications for quality control of raw intermediate and final products. The manufacturers of pharmaceuticals have to follow several international regulations like FDA 21 CFR Part 11, GMP, Gamp 5, USP<1058>, which require a lot of documentation work. The manufacturers of automatic refractometers support these users providing instrument software fulfills the requirements of 21 CFR Part 11, with user levels, electronic signature and audit trail. Furthermore, Pharma Validation and Qualification Packages are available containing

- Qualification Plan (QP)
- Design Qualification (DQ)
- Risk Analysis
- Installation Qualification (IQ)
- Operational Qualification (OQ)
- Check List 21 CFR Part 11 / SOP
- Performance Qualification (PQ)

Scales Typically Used

- Brix
- Oechsle scale
- Plato scale
- Baumé scale

Polarimeter

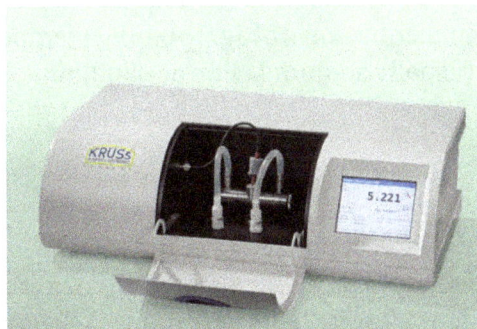

An automatic digital polarimeter

A polarimeter is a scientific instrument used to measure the angle of rotation caused by passing polarized light through an optically active substance.

Some chemical substances are optically active, and polarized (unidirectional) light will rotate either to the left (counter-clockwise) or right (clockwise) when passed through these substances. The amount by which the light is rotated is known as the angle of rotation.

History

Polarization by reflection was discovered in 1808 by Étienne-Louis Malus (1775–1812).

Measuring Principle

The ratio, the purity, and the concentration of two enantiomers can be measured via polarimetry. Enantiomers are characterized by their property to rotate the plane of linear polarized light. Therefore, those compounds are called optically active and their property is referred to as optical rotation. Light sources such as a light bulb, a light-emitting diode (LED), or the sun emit electromagnetic light waves. Their electric field oscillates in all possible planes relative to their direction of propagation. In contrast to that, the waves of linear-polarized light oscillate in parallel planes.

If light encounters a polarizer, only the part of the light that oscillates in the defined plane of the polarizer may pass through. That plane is called the plane of polarization. The plane of polarization is turned by optically active compounds. According to the direction in which the light is rotated, the enantiomer is referred to as dextrorotatory or levorotatory.

The optical activity of enantiomers is additive. If different enantiomers exist together in one solution, their optical activity adds up. That is why racemates are optically inactive, as they nullify their clockwise and counter clockwise optical activities. The optical rotation is proportional to the concentration of the optically active substances in solution. Polarimeters may therefore be applied for concentration measurements of enantiomer-pure samples. With a known concentration of a sample, polarimeters may also be applied to determine the specific rotation when characterizing a new substance. The specific rotation $[\alpha]_\lambda^T$ is a physical property and defined as the optical rotation α at a path length l of 1 dm, a concentration c of 1g/100 mL, a temperature T (usually 20 °C) and a light wavelength λ (usually sodium D line at 589.3 nm):

$$[\alpha]_\lambda^T = \frac{100 \times \alpha}{l \times c}$$

This tells us how much the plane of polarization is rotated when the ray of light passes through a specific amount of optically active molecules of a sample. Therefore, the optical rotation depends on temperature, concentration, wavelength, path length, and the substance being analyzed.

Construction

The polarimeter is made up of two Nicol prisms (the polarizer and analyzer). The polarizer is fixed and the analyzer can be rotated. The prisms may be compared to as slits S1 and S2. The light waves may be considered to correspond to waves in the string. The polarizer S1 allows only those light waves which move in a single plane. This causes the light to become plane polarized. When the analyzer is also placed in a similar position it allows the light waves coming from the polarizer to pass through it. When it is rotated through the right angle no waves can pass through the right angle and the field appears to be dark. If now a glass tube containing an optically active solution is placed between the polarizer and analyzer the light now rotates through the plane of polarization through a certain angle, the analyzer will have to be rotated in same angle.

Operation

Polarimeters measure this by passing monochromatic light through the first of two polarising plates, creating a polarized beam. This first plate is known as the polarizer. This beam is then rotated as it passes through the sample. After passing through the sample, a second polarizer, known as the analyzer, rotates either via manual rotation or automatic detection of the angle. When the analyzer is rotated to the proper angle, the maximum amount of light will pass through and shine onto a detector.

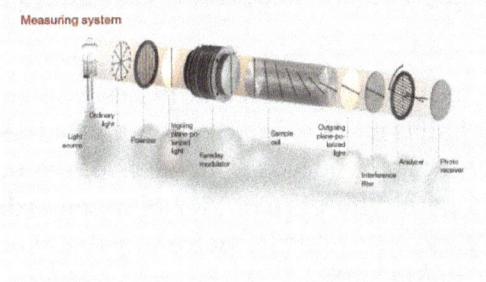

schematic set up of an automatic polarimeter: light source, polarizer, Faraday modulator, temperature controlled sample cell, interference filter of a defined wavelength, analyzer and photo receiver

Types of Polarimeter

Laurent's Half-shade Polarimeter

When plane-polarised light passes through some crystals, the velocity of left-polarised light is different from that of the right-polarised light thus the crystals are said to have two refractive indices i.e. double refracting Construction: It consists of a monochromatic source S which is placed at focal point of a convex lens L. Just after the convex lens there is a Nicol Prism P which acts as a polariser. H is a half shade device which divides the field of polarised light emerging out of the Nicol P into two halves generally of unequal brightness. T is a glass tube in which optically active solution is filled. The light after passing through T is allowed to fall on the analyzing Nicol A which can be rotated about the axis of the tube. The rotation of analyser can be measured with the help of a scale C. Laurent's half-shade polarimeter.

Working: In order to understand the need of a half-shade device, let us suppose that half-shade device is not present. The position of the analyzer is so adjusted that the field of view is dark when tube is empty. The position of the analyzer is noted on circular scale. Now the tube is filled with optically active solution and it is set in its proper position. The optically active solution rotates the plane of polarization of the light emerging out of the polariser P by some angle. So the light is transmitted by analyzer A and the field of view of telescope becomes bright. Now the analyzer is rotated by a finite angle so that the field of view of telescope again become dark. This will happen only when the analyzer is rotated by the same angle by which plane of polarization of light is rotated by optically active solution.

The position of analyzer is again noted. The difference of the two readings will give you angle of rotation of plane of polarization (8).

A difficulty is faced in the above procedure that when analyzer is rotated for the total darkness, then it is attained gradually and hence it is difficult to find the exact position correctly for which complete darkness is obtained.

To overcome above difficulty half-shade device is introduced between polariser P and glass tube T.

Half Shade Device It consist of two semicircular plates ACB and ADC. One half ACB is made of glass while other half is made of quartz. Both the halves are cemented together. The quartz is cut parallel to the optic axis. Thickness of the quartz is selected in such a way that it introduces a path difference of 'A/2 between ordinary and extraordinary ray. The thickness of the glass is selected in such a way that it absorbs the same amount of light as is absorbed by quartz half.

Let us consider that the vibration of polarisation is along OP. On passing through the glass half the vibrations remain along OP. But on passing through quartz half these vibrations will split into o- and £-components. The £-components are parallel to the optic axis while O- component is perpendicular to optic axis. The O-component travels faster in quartz and hence an emergence o-component will be along OD instead of along OC. Thus components OA and OO will combine to form a resultant vibration along OQ which makes same angle with optic axis as OP. Now if the Principal plane of the analyzing Nicol is parallel to OP then the light will pass through glass half unobstructed. Hence glass half will be brighter than quartz half or we can say that glass half will be bright and the quartz half will be dark. Similarly if principal plane of analyzing Nicol is parallel to OQ then quartz half will be bright and glass half will be dark.

When the principal plane of analyzer is along AOB then both halves will be equally bright. On the other hand if the principal plane of analyzer is along DOC. then both the halves will be equally dark.

Thus it is clear that if the analyzing Nicol is slightly disturbed from DOC then one half becomes brighter than the other. Hence by using half shade device, one can measure angle of rotation more accurately.

Determination of Specific Rotation

In order to determine specific rotation of an optically active substance (say sugar) the polarimeter tube T is first filled with pure water and analyzer is adjusted for equal darkness (Both the halves should be equally dark) point. The position of the analyzer is noted with the help of scale. Now the polarimeter tube is filled with sugar solution of known concentration and again the analyser is adjusted in such a way that again equally dark point is achieved. The position of the analyzer is again noted. The difference of the two readings will give you angle of rotation θ. Hence specific rotation S is determined by using the relation.

$$[S]t\,\lambda = \theta\,/LC$$

The above procedure may be repeated for different concentration.

Biquartz Polarimeter

In biquartz polarimeters, a biquartz plate is used. Biquartz plate consists of two semi circular plates of quartz each of thickness 3.75mm. One half consists of right-handed optically active quartz,while the other is left-handed optically active quartz.

Quartz-Wedge Polarimeter

Manual

The earliest polarimeters, which date back to the 1830s, required the user to physically rotate one polarizing element (the analyzer) whilst viewing through another static element (the detector). The detector was positioned at the opposite end of a tube containing the optically active sample, and the user used his/her eye to judge the "alignment" when least light was observed. The angle of rotation was then read from a simple protractor fixed to the moving polariser to within a degree or so.

Although most manual polarimeters produced today still adopt this basic principle, the many developments applied to the original opto-mechanical design over the years have significantly improved measurement performance. The introduction of a half-wave plate increased "distinction sensitivity", whilst a precision glass scale with vernier drum facilitated the final reading to within ca. ±0.05°. Most modern manual polarimeters also incorporate a long-life yellow LED in place of the more costly sodium arc lamp as a light source.

Semi-automatic

Today, semi-automatic polarimeters are available. The operator views the image via a digital display adjusts the analyzer angle with electronic controls.

Fully Automatic

Fully automatic polarimeters are now available and simply require the user to press a button and wait for a digital readout. Fast automatic digital polarimeters yield an accurate result within a second, regardless of the rotation angle of the sample. In addition, they provide continuous measurement, facilitating High-performance liquid chromatography and other kinetic investigations.

Another feature of modern polarimeters is the Faraday modulator. The Faraday modulator creates an alternating current magnetic field. It oscillates the plane of polarization to enhance the detection accuracy by allowing the point of maximal darkness to be passed through again and again and thus be determined with even more accuracy.

As the temperature of the sample has a significant influence on the optical rotation of the sample, modern polarimeters have already included Peltier Elements to actively control the temperature. Special techniques like a temperature controlled sample tube reduce measuring errors and ease operation. Results can directly be transferred to computers or networks for automatic processing. Traditionally, accurate filling of the sample cell had to be checked outside the instrument, as an appropriate control from within the device was not possible. Nowadays a camera system allows accurate monitoring of the sample and filling conditions in the sample cell from inside the instrument. A telecentric camera gives a sharp image over the complete length of any sample cell placed within modern instruments. The online monitoring of the filling process ensures that no bubbles or particles obstruct the measurement. A picture can be saved together with the recorded data. Any temperature gradients, inhomogeneous sample distributions or air bubbles can immediately be recognized before measurement, so that potential errors caused by bubbles or particles are no longer an issue.

Sources of Error

The angle of rotation of an optically active substance can be affected by:

- Concentration of the sample

- Wavelength of light passing through the sample (generally, angle of rotation and wavelength tend to be inversely proportional)

- Temperature of the sample (generally the two are directly proportional)

- Length of the sample cell (input by the user into most automatic polarimeters to ensure better accuracy)

- Filling conditions (bubbles, temperature and concentration gradients)

Most modern polarimeters have methods for compensating or/and controlling these errors.

Calibration

Polarimeters can be calibrated – or at least verified – by measuring a quartz plate, which is constructed to always read at a certain angle of optical rotation (usually +34°, but +17° and +8.5° are also popular depending on the sample). Quartz plates are preferred by many users because solid samples are much less affected by variations in temperature, and do not need to be mixed on-demand like sucrose solutions.

Applications

Because many optically active chemicals such as tartaric acid, are stereoisomers, a polarimeter can be used to identify which isomer is present in a sample – if it rotates polarized light to the left, it is a levo-isomer, and to the right, a dextro-isomer. It can also be used to measure the ratio of enantiomers in solutions.

The optical rotation is proportional to the concentration of the optically active substances in solution. Polarimetry may therefore be applied for concentration measurements of enantiomer-pure samples. With a known concentration of a sample, polarimetry may also be applied to determine the specific rotation (a physical property) when characterizing a new substance.

Chemical Industry

Many chemicals exhibit a specific rotation as a unique property (an intensive property like refractive index or Specific gravity) which can be used to distinguish it. Polarimeters can identify unknown samples based on this if other variables such as concentration and length of sample cell length are controlled or at least known. This is used in the chemical industry.

By the same token, if the specific rotation of a sample is already known, then the concentration and/or purity of a solution containing it can be calculated.

Most automatic polarimeters make this calculation automatically, given input on variables from the user.

Food, Beverage and Pharmaceutical Industries

Concentration and purity measurements are especially important to determine product or ingredient quality in the food & beverage and pharmaceutical industries. Samples that display specific rotations that can be calculated for purity with a polarimeter include:

- Steroids

- Diuretics

- Antibiotics

- Narcotics

- Vitamins

- Analgesics

- Amino acids

- Essential oils

- Polymers

- Starches

- Sugars

Polarimeters are used in the sugar industry for determining quality of both juice from sugar cane and the refined sucrose. Often, the sugar refineries use a modified polarimeter with a flow cell (and used in conjunction with a refractometer) called a saccharimeter. These instruments use the International Sugar Scale, as defined by the International Commission for Uniform Methods of Sugar Analysis (ICUMSA).

Spectrometer

In physics, a spectrometer is an apparatus to measure a spectrum. Generally, a spectrum is a graph that shows intensity as a function of wavelength, of frequency, of energy, of momentum, or of mass.

Optical Spectrometer

Optical spectrometers (often simply called "spectrometers"), in particular, show the intensity of light as a function of wavelength or of frequency. The deflection is produced either by refraction in a prism or by diffraction in a diffraction grating.

Mass Spectrometer

A mass spectrometer is an analytical instrument that is used to identify the amount and type of chemicals present in a sample by measuring the mass-to-charge ratio and abundance of gas-phase ions.

Time-of-flight Spectrometer

The energy spectrum of particles of known mass can also be measured by determining the time of flight between two detectors (and hence, the velocity) in a time-of-flight spectrometer. Alternatively, if the velocity is known, masses can be determined in a time-of-flight mass spectrometer.

Magnetic Spectrometer

When a fast charged particle (charge q, mass m) enters a constant magnetic field B at right angles, it is deflected into a circular path of radius r, due to the Lorentz force. The momentum p of the particle is then given by

$$p = mv = qBr,$$

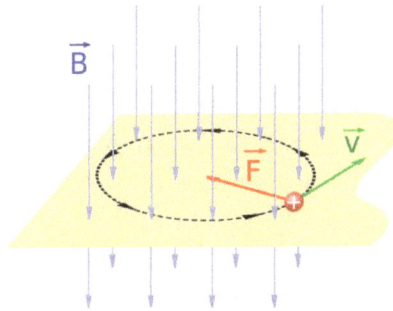

A positive charged particle moving in a circle under the influence of the Lorentz force F

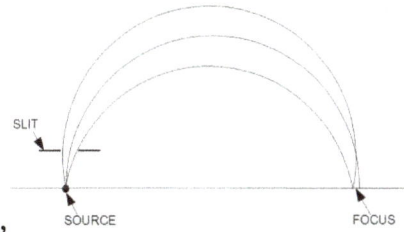

Focus of a magnetic semicircular spectrometer

where m and v are mass and velocity of the particle. The focussing principle of the oldest and simplest magnetic spectrometer, the semicircular spectrometer, invented by J. K. Danisz, is shown on the left. A constant magnetic field is perpendicular to the page. Charged particles of momentum p that pass the slit are deflected into circular paths of radius $r = p/qB$. It turns out that they all hit the horizontal line at nearly the same place, the focus; here a particle counter should be placed. Varying B, this makes possible to measure the energy spectrum of alpha particles in an alpha particle spectrometer, of beta particles in a beta particle spectrometer, of particles (e.g., fast ions) in a particle spectrometer, or to measure the relative content of the various masses in a mass spectrometer.

Since Danysz' time, many types of magnetic spectrometers more complicated than the semicircular type have been devised.

Resolution

Generally, the resolution of an instrument tells us how well two close-lying energies (or wavelengths, or frequencies, or masses) can be resolved. Generally, for an instrument with mechanical slits, higher resolution will mean lower intensity.

Autocollimator

An autocollimator is an optical instrument for non-contact measurement of angles. They are typically used to align components and measure deflections in optical or mechanical systems. An autocollimator works by projecting an image onto a target mirror and measuring the deflection of the returned image against a scale, either visually or by means of an electronic detector. A visual autocollimator can measure angles as small as 0.5 arcminute (0.15 mrad), while an electronic autocollimator can have up to 100 times more resolution.

Visual autocollimators are often used for lining up laser rod ends and checking the face parallelism of optical windows and wedges. Electronic and digiital autocollimators are used as angle measurement standards, for monitoring angular movement over long periods of time and for checking angular position repeatability in mechanical systems. Servo autocollimators are specialized compact forms of electronic autocollimators that are used in high-speed servo-feedback loops for stable-platform applications. An electronic autocollimator is typically calibrated to read the actual mirror angle.

Lensmeter

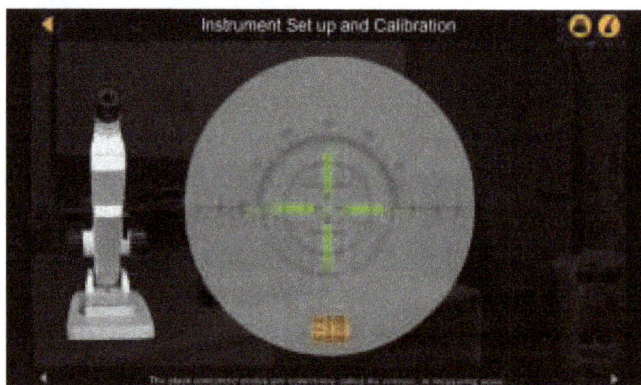

One of five comprehensive instructional videos derived from the original Flash animation on the use of the Topcon lensmeter (vertometer). Produced by Dr Suzane Vassallo with La Trobe University. These videos are also available Wikiversity, Youtube and Archive.org.

A Lensmeter in an Optical shop

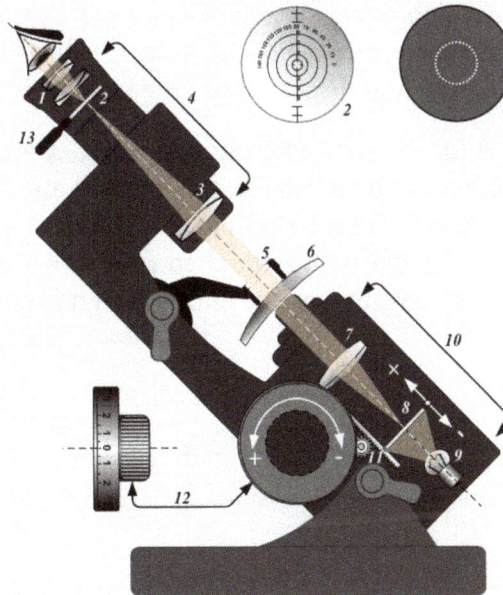

A simple lensmeter cross sectional view.

1 – Adjustable eyepiece 2 – Reticle
3 – Objective lens 4 – Keplerian telescope
5 – Lens holder 6 – Unknown lens
7 – Standard lens 8 – Illuminated target
9 – Light source 10 – Collimator
11 – Angle adjustment lever
12 – Power drum (+20 and -20 Diopters)
13 – Prism scale knob

A lensmeter or lensometer, also known as a focimeter or vertometer, is an ophthalmic instrument. It is mainly used by optometrists and opticians to verify the correct prescription in a pair of eyeglasses, to properly orient and mark uncut lenses, and to confirm the correct mounting of lenses in spectacle frames. Lensmeters can also verify the power of contact lenses, if a special lens support is used.

The parameters appraised by a lensmeter are the values specified by an ophthalmologist or optometrist on the patient's prescription: sphere, cylinder, axis, add, and in some cases, prism. The lensmeter is also used to check the accuracy of progressive lenses, and is often capable of marking the lens center and various other measurements critical to proper performance of the lens. It may also be used prior to an eye examination to obtain the last prescription the patient was given, in order to expedite the subsequent examination.

History

In 1848, Antoine Claudet produced the photographometer, an instrument designed to measure the intensity of photogenic rays; and in 1849 he brought out the focimeter, for securing a perfect focus in photographic portraiture. In 1876, Hermann Snellen introduced a phakometer which was a similar set up to an optical bench which could measure the power and find the optical centre of a convex lens. Troppman went a step further in 1912, introducing the first direct measuring instrument.

In 1922, a patent was filed for the first projection lensmeter, which has a similar system to the standard lensmeter pictured above, but projects the measuring target onto a screen eliminating the need for correction of the observer's refractive error in the instrument itself and reducing the requirement to peer down a small telescope into the instrument. Despite these advantages the above design is still predominant in the optical world.

Waveplate

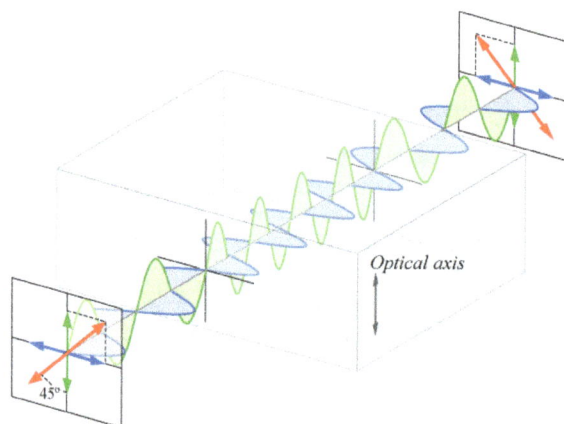

A half-wave plate. Linearly polarized light entering a waveplate can be resolved into two waves, parallel (shown as green) and perpendicular (blue) to the optical axis of the waveplate. In the plate, the parallel wave propagates slightly slower than the perpendicular one. At the far side of the plate, the parallel wave is exactly half of a wavelength delayed relative to the perpendicular wave, and the resulting combination (red) is a mirror-image of the entry polarization state (relative to the optical axis).

A waveplate or retarder is an optical device that alters the polarization state of a light wave travelling through it. Two common types of waveplates are the *half-wave plate*, which shifts the polarization direction of linearly polarized light, and the *quarter-wave plate*, which converts linearly polarized light into circularly polarized light and vice versa. A quarter wave plate can be used to produce elliptical polarization as well.

Waveplates are constructed out of a birefringent material (such as quartz or mica), for which the index of refraction is different for different orientations of light passing through it. The behavior of a waveplate (that is, whether it is a half-wave plate, a quarter-wave plate, etc.) depends on the thickness of the crystal, the wavelength of light, and the variation of the index of refraction. By appropriate choice of the relationship between these parameters, it is possible to introduce a controlled phase shift between the two polarization components of a light wave, thereby altering its polarization.

A common use of wave plates - particularly the sensitive tint (full wave) and quarter wave plates - is in optical mineralogy. Addition of plates between the polarisers of a petrographic microscope makes easier the optical identification of minerals in thin sections of rocks, in particular by allowing deduction of the shape and orientation of the optical indicatrices within the visible crystal sections. This alignment can allow discrimination between minerals which otherwise appear very similar in plane polarised and cross polarized light.

Principles of Operation

A waveplate works by shifting the phase between two perpendicular polarization components of the light wave. A typical waveplate is simply a birefringent crystal with a carefully chosen orientation and thickness. The crystal is cut into a plate, with the orientation of the cut chosen so that the optic axis of the crystal is parallel to the surfaces of the plate. This results in two axes in the plane of the cut: the *ordinary axis*, with index of refraction n_o, and the *extraordinary axis*, with index of refraction n_e. The ordinary axis is perpendicular to the optic axis. The extraordinary axis is parallel to the optic axis. For a light wave normally incident upon the plate, polarization component along the ordinary axis travels through the crystal with a speed $v_o = c/n_o$, while the polarization component along the extraordinary axis travels with a speed $v_e = c/n_e$. This leads to a phase difference between the two components as they exit the crystal. When $n_e < n_o$, as in calcite, the extraordinary axis is called the *fast axis* and the ordinary axis is called the *slow axis*. For $n_e > n_o$ the situation is reversed.

Depending on the thickness of the crystal, light with polarization components along both axes will emerge in a different polarization state. The waveplate is characterized by the amount of relative phase, Γ, that it imparts on the two components, which is related to the birefringence Δn and the thickness L of the crystal by the formula

$$\Gamma = \frac{2\pi \Delta n L}{\lambda_0},$$

where λ_0 is the vacuum wavelength of the light.

Waveplates in general as well as polarizers can be described using the Jones matrix formalism, which uses a vector to represent the polarization state of light and a matrix to represent the linear transformation of a waveplate or polarizer.

Although the birefringence Δn may vary slightly due to dispersion, this is negligible compared to the variation in phase difference according to the wavelength of the light due to the fixed path difference (λ_0 in the denominator in the above equation). Waveplates are thus manufactured to work for a particular range of wavelengths. The phase variation can be minimized by stacking two waveplates that differ by a tiny amount in thickness back-to-back, with the slow axis of one along the fast axis of the other. With this configuration, the relative phase imparted can be, for the case of a quarter-wave plate, one-fourth a wavelength rather than three-fourths or one-fourth plus an integer. This is called a *zero-order waveplate*.

For a single waveplate changing the wavelength of the light introduces a linear error in the phase. Tilt of the waveplate enters via a factor of $1/\cos\theta$ (where θ is the angle of tilt) into the path length and thus only quadratically into the phase. For the extraordinary polarization the tilt also changes the refractive index to the ordinary via a factor of $\cos\theta$, so combined with the path length, the phase shift for the extraordinary light due to tilt is zero.

A polarization-independent phase shift of zero order needs a plate with thickness of one wavelength. For calcite the refractive index changes in the first decimal place, so that a true zero order plate is ten times as thick as one wavelength. For quartz and magnesium fluoride the refractive index changes in the second decimal place and true zero order plates are common for wavelengths above 1 μm.

Plate Types

Half-wave Plate

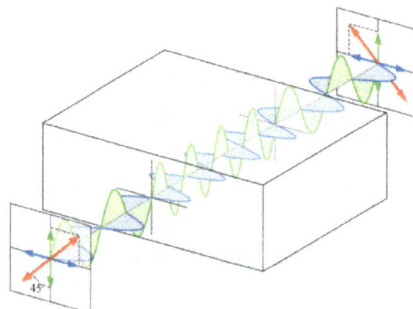

A wave passing through a half-wave plate.

For a half-wave plate, the relationship between L, Δn, and λ_0 is chosen so that the phase shift between polarization components is $\Gamma = \pi$. Now suppose a linearly polarized wave with polarization vector $\hat{\mathbf{p}}$ is incident on the crystal. Let θ denote the angle between $\hat{\mathbf{p}}$ and $\hat{\mathbf{f}}$, where $\hat{\mathbf{f}}$ is the vector along the waveplate's fast axis. Let z denote the propagation axis of the wave. The electric field of the incident wave is

$$\mathbf{E}e^{i(kz-\omega t)} = E\hat{\mathbf{p}}e^{i(kz-\omega t)} = E(\cos\theta\hat{\mathbf{f}} + \sin\theta\hat{\mathbf{s}})e^{i(kz-\omega t)},$$

where $\hat{\mathbf{s}}$ lies along the waveplate's slow axis. The effect of the half-wave plate is to introduce a phase shift term $e^{i\Gamma} = e^{i\pi} = -1$ between the f and s components of the wave, so that upon exiting the crystal the wave is now given by

$$E(\cos\theta\hat{\mathbf{f}} - \sin\theta\hat{\mathbf{s}})e^{i(kz-\omega t)} = E[\cos(-\theta)\hat{\mathbf{f}} + \sin(-\theta)\hat{\mathbf{s}}]e^{i(kz-\omega t)}.$$

If $\hat{\mathbf{p}}'$ denotes the polarization vector of the wave exiting the waveplate, then this expression shows that the angle between $\hat{\mathbf{p}}'$ and $\hat{\mathbf{f}}$ is $-\theta$. Evidently, the effect of the half-wave plate is to mirror the wave's polarization vector through the plane formed by the vectors $\hat{\mathbf{f}}$ and $\hat{\mathbf{z}}$. For linearly polarized light, this is equivalent to saying that the effect of the half-wave plate is to rotate the polarization vector through an angle 2θ; however, for elliptically polarized light the half-wave plate also has the effect of inverting the light's handedness.

Quarter-wave Plate

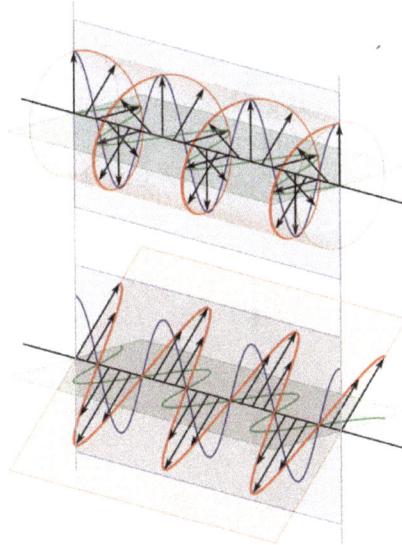

Two waves differing by a quarter-phase shift for one axis.

Creating circular polarization using a quarter-wave plate and a polarizing filter

For a quarter-wave plate, the relationship between L, Δn, and λ_0 is chosen so that the phase shift between polarization components is $\Gamma = \pi/2$. Now suppose a linearly polarized wave is incident on the crystal. This wave can be written as

$$(E_f\hat{\mathbf{f}} + E_s\hat{\mathbf{s}})e^{i(kz-\omega t)},$$

where the f and s axes are the quarter-wave plate's fast and slow axes, respectively, the wave propagates along the z axis, and E_f and E_s are real. The effect of the quarter-wave plate is to introduce a phase shift term $e^{i\Gamma} = e^{i\pi/2} = i$ between the f and s components of the wave, so that upon exiting the crystal the wave is now given by

$$(E_f\hat{\mathbf{f}} + iE_s\hat{\mathbf{s}})e^{i(kz-\omega t)}.$$

The wave is now elliptically polarized.

If the axis of polarization of the incident wave is chosen so that it makes a 45° with the fast and slow axes of the waveplate, then $E_f = E_s \equiv E$, and the resulting wave upon exiting the waveplate is

$$E(\hat{\mathbf{f}} + i\hat{\mathbf{s}})e^{i(kz-\omega t)},$$

and the wave is circularly polarized.

If the axis of polarization of the incident wave is chosen so that it makes a 0° with the fast or slow axes of the waveplate, then the polarization will not change, so remains linear. If the angle is in between 0° and 45° the resulting wave has an elliptical polarization.

A circulating polarization looks strange, but can be easier imagined as the sum of two linear polarizations with a phase difference of 90°. The output depends on the polarization of the input. Suppose polarization axes x and y parallel with the fast and slow axis of the wave plate:

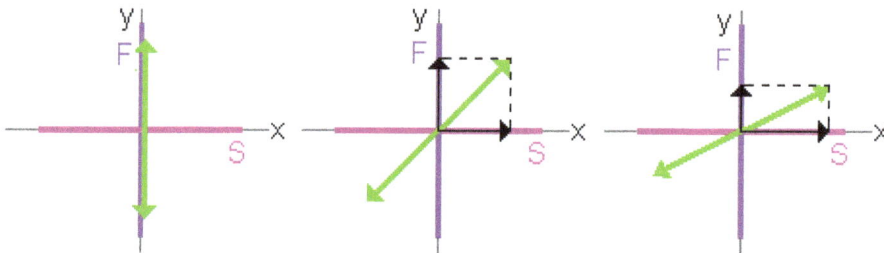

The polarization of the incoming photon (or beam) can be resolved as two polarizations on the x and y axis. If the input polarization is parallel to the fast or slow axis, then there is no polarization of the other axis, so the output polarization is the same as the input (only the phase more or less delayed). If the input polarization is 45° to the fast and slow axis, the polarization on those axes are equal. But the phase of the output of the slow axis will be delayed 90° with the output of the fast axis. If not the amplitude but both sinus values are displayed, then x and y combined will describe a circle. With other angles than 0° or 45° the values in fast and slow axis will differ and their resultant output will describe an ellipse.

Full Wave, or Sensitive Tint Plate

A full wave plate introduces a phase difference between the two polarisation directions of exactly one wavelength for green light (wavelength = 540 nm). This means that white light which passes through the plate but is then linearly polarised will have this green wavelength fully extinguished, but still retain elements of other colors. This means that under these conditions the plate will appear an intense shade of red-violet, sometimes known as sensitive tint. This gives rise to this plate's alternative names, the *sensitive tint plate* or (less commonly) *red tint plate*. These plates are widely used in optical mineralogy to make easier the optical identification of minerals in thin sections of rocks.

Use of Waveplates in Mineralogy and Optical Petrology

The sensitive tint (full wave) and quarter wave plates are widely used in the field of optical mineralogy. Addition of plates between the polarisers of a petrographic microscope makes easier the optical identification of minerals in thin sections of rocks, in particular by allowing deduction of the shape and orientation of the optical indicatrices within the visible crystal sections.

In practical terms, the plate is inserted between the perpendicular polarisers at an angle of 45 degrees. This allows two different procedures to be carried out to investigate the mineral under the crosshairs of the microscope. More simply, in ordinary cross polarised light, the plate can be used to distinguish the orientation of the optical indicatrix relative to crystal elongation - that is, whether the mineral is "length slow" or "length fast" - based on whether the visible interference colours increase or decrease by one order when the plate is added. A slightly more complex procedure allows for a tint plate to be used in conjunction with interference figure techniques to allow measurement of the *optic angle* of the mineral. The optic angle (often notated as "2V") can both be diagnostic of mineral type, as well as in some cases revealing information about the variation of chemical composition within a single mineral type.

References

- Brians, Paul (2003). Common Errors in English. Franklin, Beedle & Associates. p. 125. ISBN 1-887902-89-9.

- Tilton, Buck (2005). The Complete Book of Fire: Building Campfires for Warmth, Light, Cooking, and Survival. Menasha Ridge Press. p. 25. ISBN 0-89732-633-4.

- Glick, Thomas F.; Steven John Livesey; Faith Wallis (2005). Medieval science, technology, and medicine: an encyclopedia. Routledge. p. 167. ISBN 978-0-415-96930-7. Retrieved 24 April 2011.

- Henry C. King (28 September 2003). The History of the Telescope. Courier Dover Publications. p. 27. ISBN 978-0-486-43265-6. Retrieved 6 June 2012.

- Paul S. Agutter; Denys N. Wheatley (12 December 2008). Thinking about Life: The History and Philosophy of Biology and Other Sciences. Springer. p. 17. ISBN 978-1-4020-8865-0. Retrieved 6 June 2012.

- Vincent Ilardi (2007). Renaissance Vision from Spectacles to Telescopes. American Philosophical Society. p. 210. ISBN 978-0-87169-259-7. Retrieved 6 June 2012.

- Fred Watson (1 October 2007). Stargazer: The Life and Times of the Telescope. Allen & Unwin. p. 55. ISBN 978-1-74175-383-7. Retrieved 6 June 2012.

- Gould, Stephen Jay (2000). "Chapter 2: The Sharp-Eyed Lynx, Outfoxed by Nature". The Lying Stones of Marrakech: Penultimate Reflections in Natural History. New York, N.Y: Harmony. ISBN 0-224-05044-3.

- Wootton, David (2006). Bad medicine: doctors doing harm since Hippocrates. Oxford [Oxfordshire]: Oxford University Press. ISBN 0-19-280355-7

- Morita S (2006). Roadmap of Scanning Probe Microscopy. NanoScience and Technology. Berlin: Springer. ISBN 3-540-34314-8.[

- Stephen G. Lipson, Ariel Lipson, Henry Lipson, Optical Physics 4th Edition, Cambridge University Press, ISBN 978-0-521-49345-1

Optical Communication: An Overview

Communication that takes places at a distance by using light to carry information is optical communication while the method that is required to transmit information by light through an optical fiber is fiber optic communication. The aspects elucidated in this chapter are of vital importance for the progress of this field and its allied branches.

Optical Communication

A naval signal lamp, a form of optical communication that uses shutters and is typically employed with Morse code (2002)

Optical communication, also known as optical telecommunication, is communication at a distance using light to carry information. It can be performed visually or by using electronic devices. The earliest basic forms of optical communication date back several millennia, while the earliest electrical device created to do so was the photophone, invented in 1880.

An optical communication system uses a transmitter, which encodes a message into an optical signal, a channel, which carries the signal to its destination, and a receiver, which reproduces the message from the received optical signal. When electronic equipment is not employed the 'receiver' is a person visually observing and interpreting a signal, which may be either simple (such as the presence of a beacon fire) or complex (such as lights using color codes or flashed in a Morse code sequence).

Free-space optical communication has been deployed in space, while terrestrial forms are naturally limited by geography, weather and the availability of light. This article provides a basic introduction to different forms of optical communication.

Forms

Visual techniques such as smoke signals, beacon fires, hydraulic telegraphs, ship flags and semaphore lines were the earliest forms of optical communication. Hydraulic telegraph semaphores date back to the 4th century BCE Greece. Distress flares are still used by mariners in emergencies, while lighthouses and navigation lights are used to communicate navigation hazards.

The heliograph uses a mirror to reflect sunlight to a distant observer. When a signaler tilts the mirror to reflect sunlight, the distant observer sees flashes of light that can be used to transmit a prearranged signaling code. Naval ships often use signal lamps and Morse code in a similar way.

Aircraft pilots often use visual approach slope indicator (VASI) projected light systems to land safely, especially at night. Military aircraft landing on an aircraft carrier use a similar system to land correctly on a carrier deck. The coloured light system communicates the aircraft's height relative to a standard landing glideslope. As well, airport control towers still use Aldis lamps to transmit instructions to aircraft whose radios have failed.

In the present day a variety of electronic systems optically transmit and receive information carried by pulses of light. Fiber-optic communication cables are now employed to send the great majority of the electronic data and long distance telephone calls that are not conveyed by either radio, terrestrial microwave or satellite. Free-space optical communications are also used every day in various applications.

Semaphore Line

A replica of one of Chappe's semaphore towers (18th century).

A 'semaphore telegraph', also called a 'semaphore line', 'optical telegraph', 'shutter telegraph chain', 'Chappe telegraph', or 'Napoleonic semaphore', is a system used for conveying information by means of visual signals, using towers with pivoting arms or shutters, also known as blades or paddles. Information is encoded by the position of the mechanical elements; it is read when the shutter is in a fixed position.

Semaphore lines were a precursor of the electrical telegraph. They were far faster than post riders for conveying a message over long distances, but far more expensive and less private than the electrical telegraph lines which would later replace them. The maximum distance that a pair of semaphore telegraph stations can bridge is limited by geography, weather and the availability of light; thus, in practical use, most optical telegraphs used lines of relay stations to bridge longer distances. Each relay station would also require its complement of skilled operator-observers to convey messages back and forth across the line.

The modern design of semaphores was first foreseen by the British polymath Robert Hooke, who first gave a vivid and comprehensive outline of visual telegraphy in an 1684 submission to the Royal Society. His proposal (which was motivated by military concerns following the Battle of Vienna the preceding year) was not put into practice during his lifetime.

The first operational optical semaphore line arrived in 1792, created by the French engineer Claude Chappe and his brothers, who succeeded in covering France with a network of 556 stations stretching a total distance of 4,800 kilometres (3,000 mi). It was used for military and national communications until the 1850s.

Many national services adopted signaling systems different from the Chappe system. For example, Britain and Sweden adopted systems of shuttered panels (in contradiction to the Chappe brothers' contention that angled rods are more visible). In Spain, the engineer Agustín de Betancourt developed his own system which was adopted by that state. This system was considered by many experts in Europe better than Chappe's, even in France.

These systems were popular in the late 18th to early 19th century but could not compete with the electrical telegraph, and went completely out of service by 1880.

Semaphore Signal Flags

A naval signaler transmitting a message by flag semaphore (2002).

Semaphore Flags is the system for conveying information at a distance by means of visual signals with hand-held flags, rods, disks, paddles, or occasionally bare or gloved hands. Information is encoded by the position of the flags, objects or arms; it is read when they are in a fixed position.

Semaphores were adopted and widely used (with hand-held flags replacing the mechanical arms of shutter semaphores) in the maritime world in the 19th century.

They are still used during underway replenishment at sea and are acceptable for emergency communication in daylight or, using lighted wands instead of flags, at night.

The newer flag semaphore system uses two short poles with square flags, which a signaler holds in different positions to convey letters of the alphabet and numbers. The transmitter holds one pole in each hand, and extends each arm in one of eight possible directions. Except for in the rest position, the flags cannot overlap. The flags are colored differently based on whether the signals are sent by sea or by land. At sea, the flags are colored red and yellow (the Oscar flags), while on land, they are white and blue (the Papa flags). Flags are not required, they just make the characters more obvious.

Optical Fiber

Optical fiber is the most common type of channel for optical communications. The transmitters in optical fiber links are generally light-emitting diodes (LEDs) or laser diodes. Infrared light, rather than visible light is used more commonly, because optical fibers transmit infrared wavelengths with less attenuation and dispersion. The signal encoding is typically simple intensity modulation, although historically optical phase and frequency modulation have been demonstrated in the lab. The need for periodic signal regeneration was largely superseded by the introduction of the erbium-doped fiber amplifier, which extended link distances at significantly lower cost.

Signal Lamps

An air traffic controller holding a signal light gun that can be used to direct aircraft experiencing a radio failure (2007).

Signal lamps (such as Aldis lamps), are visual signaling devices for optical communication (typically using Morse code). Modern signal lamps are a focused lamp which can produce a pulse of light. In large versions this pulse is achieved by opening and closing shutters mounted in front of the lamp, either via a manually operated pressure switch or, in later versions, automatically.

With hand held lamps, a concave mirror is tilted by a trigger to focus the light into pulses. The lamps are usually equipped with some form of optical sight, and are most commonly deployed on naval vessels and also used in airport control towers with coded aviation light signals.

Aviation light signals are used in the case of a radio failure, an aircraft not equipped with a radio, or in the case of a hearing-impaired pilot. Air traffic controlers have long used signal light guns to direct such aircraft. The light gun's lamp has a focused bright beam capable of emitting three different colors: red, white and green. These colors may be flashing or steady, and provide different instructions to aircraft in flight or on the ground (for example, "cleared to land" or "cleared for takeoff"). Pilots can acknowledge the instructions by wiggling their plane's wings, moving their ailerons if they are on the ground, or by flashing their landing or navigation lights during night time. Only 12 simple standardized instructions are directed at aircraft using signal light guns as the system is not utilized with Morse code.

Photophone

The photophone (originally given an alternate name, radiophone) is a communication device which allowed for the transmission of speech on a beam of light. It was invented jointly by Alexander Graham Bell and his assistant Charles Sumner Tainter on February 19, 1880, at Bell's 1325 'L' Street laboratory in Washington, D.C. Both were later to become full associates in the Volta Laboratory Association, created and financed by Bell.

On June 21, 1880, Bell's assistant transmitted a wireless voice telephone message of considerable distance, from the roof of the Franklin School to the window of Bell's laboratory, some 213 meters (about 700 ft.) away.

Bell believed the photophone was his most important invention. Of the 18 patents granted in Bell's name alone, and the 12 he shared with his collaborators, four were for the photophone, which Bell referred to as his *'greatest achievement'*, telling a reporter shortly before his death that the photophone was *"the greatest invention [I have] ever made, greater than the telephone"*.

The photophone was a precursor to the fiber-optic communication systems which achieved popular worldwide usage starting in the 1980s. The master patent for the photophone (U.S. Patent 235,199 *Apparatus for Signalling and Communicating, called Photophone*), was issued in December 1880, many decades before its principles came to have practical applications.

Free-space Optical Communication

Free-space optics (FSO) systems are generally employed for 'last mile' telecommunications and can function over distances of several kilometers as long as there is a clear line of sight between the source and the destination, and the optical receiver can reliably decode the transmitted information. Other free-space systems can provide high-data-rate, long-range links using small, low-mass, low-power-consumption subsystems.

More generally, transmission of unguided optical signals is known as optical wireless communications (OWC). Examples include medium-range visible light communication and short-distance IrDA, using infrared LEDs.

Heliograph

Heliograph: Australians using a heliograph in North Africa (1940).

A heliograph is a wireless solar telegraph that signals by flashes of sunlight (generally using Morse code) reflected by a mirror. The flashes are produced by momentarily pivoting the mirror, or by interrupting the beam with a shutter.

The heliograph was a simple but effective instrument for instantaneous optical communication over long distances during the late 19th and early 20th century. Its main uses were in military, surveys and forest protection work. They were standard issue in the British and Australian armies until the 1960s, and were used by the Pakistani army as late as 1975.

Optical Fiber

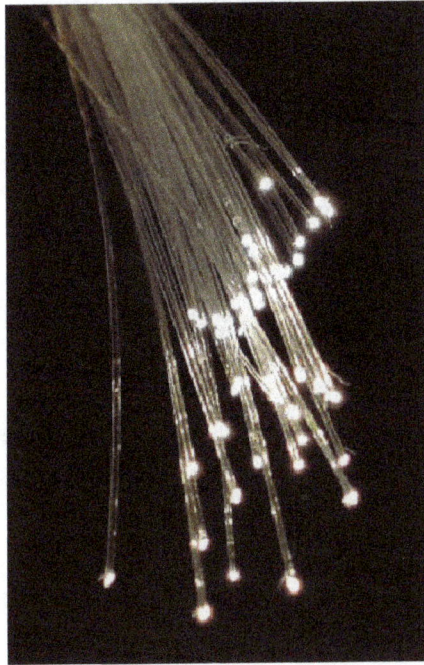

A bundle of optical fibers

Fiber crew installing a 432-count fiber cable underneath the streets of Midtown
Manhattan, New York City

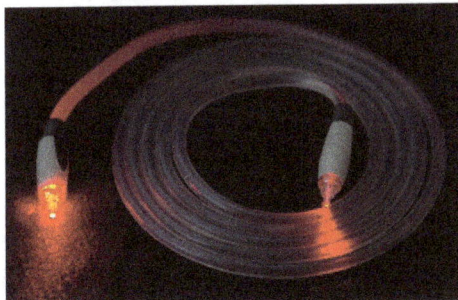

A TOSLINK fiber optic audio cable with red light being shone in one end transmits
the light to the other end

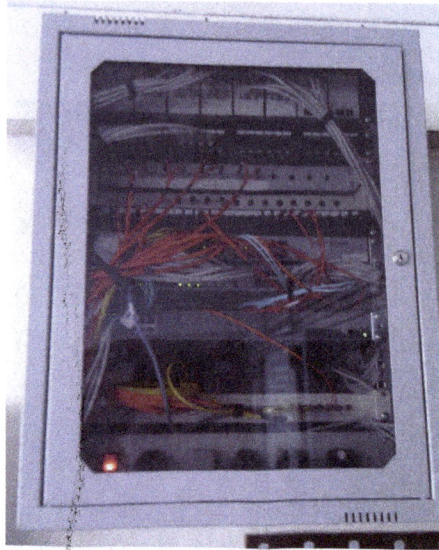

A wall-mount cabinet containing optical fiber interconnects. The yellow cables are single mode fibers; the orange and aqua cables are multi-mode fibers: 50/125 μm OM2 and 50/125 μm OM3 fibers respectively.

An optical fiber (or optical fibre) is a flexible, transparent fiber made by drawing glass (silica) or plastic to a diameter slightly thicker than that of a human hair. Optical fibers are used most often as a means to transmit light between the two ends of the fiber and find wide usage in fiber-optic communications, where they permit transmission over longer distances and at higher bandwidths (data rates) than wire cables. Fibers are used instead of metal wires because signals travel along them with lesser amounts of loss; in addition, fibers are also immune to electromagnetic interference, a problem from which metal wires suffer excessively. Fibers are also used for illumination, and are wrapped in bundles so that they may be used to carry images, thus allowing viewing in confined spaces, as in the case of a fiberscope. Specially designed fibers are also used for a variety of other applications, some of them being fiber optic sensors and fiber lasers.

Optical fibers typically include a transparent core surrounded by a transparent cladding material with a lower index of refraction. Light is kept in the core by the phenomenon of total internal reflection which causes the fiber to act as a waveguide. Fibers that support many propagation paths or transverse modes are called multi-mode fibers (MMF), while those that support a single mode are called single-mode fibers (SMF). Multi-mode fibers generally have a wider core diameter and are used for short-distance communication links and for applications where high power must be transmitted. Single-mode fibers are used for most communication links longer than 1,000 meters (3,300 ft).

An important aspect of a fiber optic communication is that of extension of the fiber optic cables such that the losses brought about by joining two different cables is kept to a minimum. Joining lengths of optical fiber often proves to be more complex than joining electrical wire or cable and involves careful cleaving of the fibers, perfect alignment of the fiber cores, and the splicing of these aligned fiber cores. For applications that demand a permanent connection a mechanical splice which holds the ends of the fibers together mechanically could be used or a fusion splice that uses heat to fuse the ends of the fibers together could be used. Temporary or semi-permanent connections are made by means of specialized optical fiber connectors.

The field of applied science and engineering concerned with the design and application of optical fibers is known as fiber optics.

History

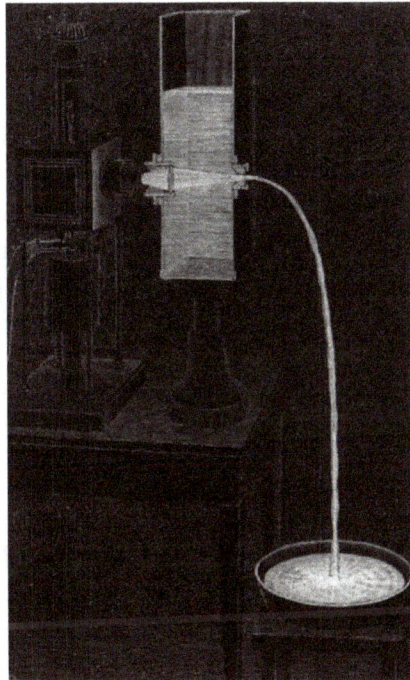

Daniel Colladon first described this "light fountain" or "light pipe" in an 1842 article titled On the reflections of a ray of light inside a parabolic liquid stream. This particular illustration comes from a later article by Colladon, in 1884.

Guiding of light by refraction, the principle that makes fiber optics possible, was first demonstrated by Daniel Colladon and Jacques Babinet in Paris in the early 1840s. John Tyndall included a demonstration of it in his public lectures in London, 12 years later. Tyndall also wrote about the property of total internal reflection in an introductory book about the nature of light in 1870:

When the light passes from air into water, the refracted ray is bent *towards* the perpendicular... When the ray passes from water to air it is bent *from* the perpendicular... If the angle which the ray in water encloses with the perpendicular to the surface be greater than 48 degrees, the ray will not quit the water at all: it will be *totally reflected* at the surface.... The angle which marks the limit where total reflection begins is called the limiting angle of the medium. For water this angle is 48°27′, for flint glass it is 38°41′, while for diamond it is 23°42′.

Unpigmented human hairs have also been shown to act as an optical fiber.

Practical applications, such as close internal illumination during dentistry, appeared early in the twentieth century. Image transmission through tubes was demonstrated independently by the radio experimenter Clarence Hansell and the television pioneer John Logie Baird in the 1920s. The principle was first used for internal medical examinations by Heinrich Lamm in the following decade. Modern optical fibers, where the glass fiber is coated with a transparent cladding to offer a more suitable refractive index, appeared later in the decade. Development then focused on fiber bundles for image transmission. Harold Hopkins and Narinder Singh Kapany at Imperial College in London achieved low-loss light transmission through a 75 cm long bundle which combined several thousand fibers. Their article titled "A flexible fibrescope, using static scanning" was published in the journal *Nature* in 1954. The first fiber optic semi-flexible gastroscope was patented by Basil Hirschowitz, C. Wilbur Peters, and Lawrence E. Curtiss, researchers at the University of Michigan, in 1956. In the process of developing the gastroscope, Curtiss produced the first glass-clad fibers; previous optical fibers had relied on air or impractical oils and waxes as the low-index cladding material.

A variety of other image transmission applications soon followed.

In 1880 Alexander Graham Bell and Sumner Tainter invented the Photophone at the Volta Laboratory in Washington, D.C., to transmit voice signals over an optical beam. It was an advanced form of telecommunications, but subject to atmospheric interferences and impractical until the secure transport of light that would be offered by fiber-optical systems. In the late 19th and early 20th centuries, light was guided through bent glass rods to illuminate body cavities. Jun-ichi Nishizawa, a Japanese scientist at Tohoku University, also proposed the use of optical fibers for communications in 1963, as stated in his book published in 2004 in India. Nishizawa invented other technologies that contributed to the development of optical fiber communications, such as the graded-index optical fiber as a channel for transmitting light from semiconductor lasers. The first working fiber-optical data transmission system was demonstrated by German physicist Manfred Börner at Telefunken Research Labs in Ulm in 1965, which was followed by the first patent application for this technology in 1966. NASA used fiber optics in the television cameras that were sent to the moon. At the time, the use in the cameras was classified *confidential*, and employees handling the cameras had to be supervised by someone with an appropriate security clearance.

Charles K. Kao and George A. Hockham of the British company Standard Telephones and Cables (STC) were the first to promote the idea that the attenuation in optical fibers could be reduced below 20 decibels per kilometer (dB/km), making fibers a practical communication medium. They proposed that the attenuation in fibers available at the time was caused by impurities that could be removed, rather than by fundamental physical effects such as scattering. They correctly and systematically theorized the light-loss properties for optical fiber, and pointed out the right material to use for such fibers — silica glass with high purity. This discovery earned Kao the Nobel Prize in Physics in 2009.

The crucial attenuation limit of 20 dB/km was first achieved in 1970, by researchers Robert D. Maurer, Donald Keck, Peter C. Schultz, and Frank Zimar working for American glass maker Corning Glass Works, now Corning Incorporated. They demonstrated a fiber with 17 dB/km attenuation by doping silica glass with titanium. A few years later they produced a fiber with only 4 dB/km attenuation using germanium dioxide as the core dopant. Such low attenuation ushered in the era of optical fiber telecommunication. In 1981, General Electric produced fused quartz ingots that could be drawn into strands 25 miles (40 km) long.

The Italian research center CSELT worked with Corning to develop practical optical fiber cables, resulting in the first metropolitan fiber optic cable being deployed in Torino in 1977. CSELT also developed an early technique for splicing optical fibers, called Springroove.

Attenuation in modern optical cables is far less than in electrical copper cables, leading to long-haul fiber connections with repeater distances of 70–150 kilometers (43–93 mi). The erbium-doped fiber amplifier, which reduced the cost of long-distance fiber systems by reducing or eliminating optical-electrical-optical repeaters, was co-developed by teams led by David N. Payne of the University of Southampton and Emmanuel Desurvire at Bell Labs in 1986. Robust modern optical fiber uses glass for both core and sheath, and is therefore less prone to aging. It was invented by Gerhard Bernsee of Schott Glass in Germany in 1973.

The emerging field of photonic crystals led to the development in 1991 of photonic-crystal fiber, which guides light by diffraction from a periodic structure, rather than by total internal reflection. The first photonic crystal fibers became commercially available in 2000. Photonic crystal fibers can carry higher power than conventional fibers and their wavelength-dependent properties can be manipulated to improve performance.

Uses

Communication

Optical fiber can be used as a medium for telecommunication and computer networking because it is flexible and can be bundled as cables. It is especially advantageous for long-distance communications, because light propagates through the fiber with little attenuation compared to electrical cables. This allows long distances to be spanned with few repeaters.

The per-channel light signals propagating in the fiber have been modulated at rates as high as 111 gigabits per second (Gbit/s) by NTT, although 10 or 40 Gbit/s is typical in deployed systems. In June 2013, researchers demonstrated transmission of 400 Gbit/s over a single channel using 4-mode orbital angular momentum multiplexing.

Each fiber can carry many independent channels, each using a different wavelength of light (wavelength-division multiplexing (WDM)). The net data rate (data rate without overhead bytes) per fiber is the per-channel data rate reduced by the FEC overhead, multiplied by the number of channels (usually up to eighty in commercial dense WDM systems as of 2008). As of 2011 the record for bandwidth on a single core was 101 Tbit/s (370 channels at 273 Gbit/s each). The record for a multi-core fiber as of January 2013 was 1.05 petabits per second. In 2009, Bell Labs broke the 100 (petabit per second)×kilometer barrier (15.5 Tbit/s over a single 7,000 km fiber).

For short distance application, such as a network in an office building, fiber-optic cabling can save space in cable ducts. This is because a single fiber can carry much more data than electrical cables such as standard category 5 Ethernet cabling, which typically runs at 100 Mbit/s or 1 Gbit/s speeds. Fiber is also immune to electrical interference; there is no cross-talk between signals in different cables, and no pickup of environmental noise. Non-armored fiber cables do not conduct electricity, which makes fiber a good solution for protecting communications equipment in high voltage environments, such as power generation facilities, or metal communication structures prone to lightning strikes. They can also be used in environments where explosive fumes are present, without danger of ignition. Wiretapping (in this case, fiber tapping) is more difficult compared to electrical connections, and there are concentric dual-core fibers that are said to be tap-proof.

Fibers are often also used for short-distance connections between devices. For example, most high-definition televisions offer a digital audio optical connection. This allows the streaming of audio over light, using the TOSLINK protocol.

Advantages Over Copper Wiring

The advantages of optical fiber communication with respect to copper wire systems are:

Broad bandwidth

> A single optical fiber can carry over 3,000,000 full-duplex voice calls or 90,000 TV channels.

Immunity to electromagnetic interference

> Light transmission through optical fibers is unaffected by other electromagnetic radiation nearby. The optical fiber is electrically non-conductive, so it does not act as an antenna to pick up electromagnetic signals. Information traveling

inside the optical fiber is immune to electromagnetic interference, even electromagnetic pulses generated by nuclear devices.

Low attenuation loss over long distances

Attenuation loss can be as low as 0.2 dB/km in optical fiber cables, allowing transmission over long distances without the need for repeaters.

Electrical insulator

Optical fibers do not conduct electricity, preventing problems with ground loops and conduction of lightning. Optical fibers can be strung on poles alongside high voltage power cables.

Material cost and theft prevention

Conventional cable systems use large amounts of copper. Global copper prices experienced a boom in the 2000s, and copper has been a target of metal theft.

Security of information passed down the cable

Copper can be tapped with very little chance of detection.

Sensors

Fibers have many uses in remote sensing. In some applications, the sensor is itself an optical fiber. In other cases, fiber is used to connect a non-fiberoptic sensor to a measurement system. Depending on the application, fiber may be used because of its small size, or the fact that no electrical power is needed at the remote location, or because many sensors can be multiplexed along the length of a fiber by using different wavelengths of light for each sensor, or by sensing the time delay as light passes along the fiber through each sensor. Time delay can be determined using a device such as an *optical time-domain reflectometer*.

Optical fibers can be used as sensors to measure strain, temperature, pressure and other quantities by modifying a fiber so that the property to measure modulates the intensity, phase, polarization, wavelength, or transit time of light in the fiber. Sensors that vary the intensity of light are the simplest, since only a simple source and detector are required. A particularly useful feature of such fiber optic sensors is that they can, if required, provide distributed sensing over distances of up to one meter. In contrast, highly localized measurements can be provided by integrating miniaturized sensing elements with the tip of the fiber. These can be implemented by various micro- and nanofabrication technologies, such that they do not exceed the microscopic boundary of the fiber tip, allowing such applications as insertion into blood vessels via hypodermic needle.

Extrinsic fiber optic sensors use an optical fiber cable, normally a multi-mode one, to transmit modulated light from either a non-fiber optical sensor—or an electronic sensor connected to an optical transmitter. A major benefit of extrinsic sensors is their ability to reach otherwise inaccessible places. An example is the measurement of temperature inside aircraft jet engines by using a fiber to transmit radiation into a radiation pyrometer outside the engine. Extrinsic sensors can be used in the same way to measure the internal temperature of electrical transformers, where the extreme electromagnetic fields present make other measurement techniques impossible. Extrinsic sensors measure vibration, rotation, displacement, velocity, acceleration, torque, and twisting. A solid state version of the gyroscope, using the interference of light, has been developed. The *fiber optic gyroscope (FOG)* has no moving parts, and exploits the *Sagnac effect* to detect mechanical rotation.

Common uses for fiber optic sensors includes advanced intrusion detection security systems. The light is transmitted along a fiber optic sensor cable placed on a fence, pipeline, or communication cabling, and the returned signal is monitored and analyzed for disturbances. This return signal is digitally processed to detect disturbances and trip an alarm if an intrusion has occurred.

Optical fibers are widely used as components of optical chemical sensors and optical biosensors.

Power Transmission

Optical fiber can be used to transmit power using a photovoltaic cell to convert the light into electricity. While this method of power transmission is not as efficient as conventional ones, it is especially useful in situations where it is desirable not to have a metallic conductor as in the case of use near MRI machines, which produce strong magnetic fields. Other examples are for powering electronics in high-powered antenna elements and measurement devices used in high-voltage transmission equipment.

Other Uses

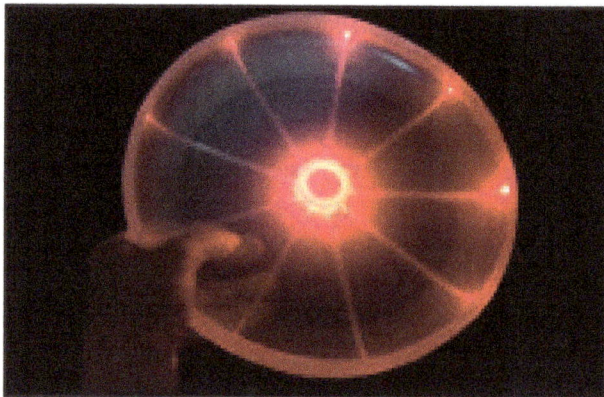

A frisbee illuminated by fiber optics

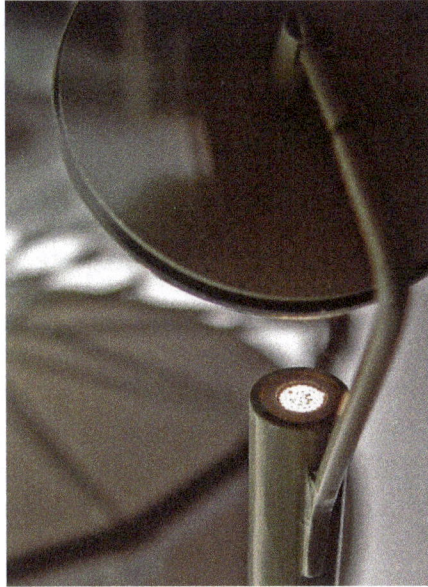

Light reflected from optical fiber illuminates exhibited model

Optical fibers have a wide number of applications. They are used as light guides in medical and other applications where bright light needs to be shone on a target without a clear line-of-sight path. In some buildings, optical fibers route sunlight from the roof to other parts of the building. Optical fiber lamps are used for illumination in decorative applications, including signs, art, toys and artificial Christmas trees. Swarovski boutiques use optical fibers to illuminate their crystal showcases from many different angles while only employing one light source. Optical fiber is an intrinsic part of the light-transmitting concrete building product, LiTraCon.

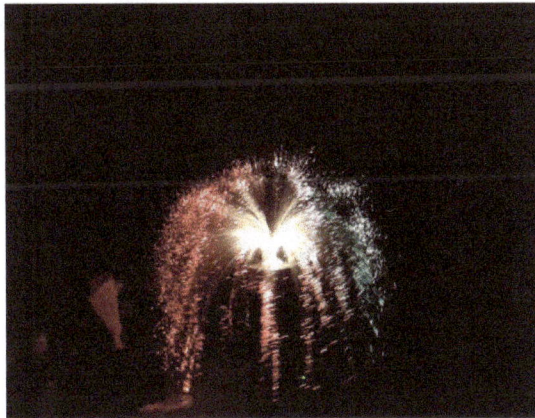

Use of optical fiber in a decorative lamp or nightlight.

Optical fiber is also used in imaging optics. A coherent bundle of fibers is used, sometimes along with lenses, for a long, thin imaging device called an endoscope, which is used to view objects through a small hole. Medical endoscopes are used for minimally invasive exploratory or surgical procedures. Industrial endoscopes are used for inspecting anything hard to reach, such as jet engine interiors. Many microscopes use fiber-optic light sources to provide intense illumination of samples being studied.

In spectroscopy, optical fiber bundles transmit light from a spectrometer to a substance that cannot be placed inside the spectrometer itself, in order to analyze its composition. A spectrometer analyzes substances by bouncing light off and through them. By using fibers, a spectrometer can be used to study objects remotely.

An optical fiber doped with certain rare earth elements such as erbium can be used as the gain medium of a laser or optical amplifier. Rare-earth-doped optical fibers can be used to provide signal amplification by splicing a short section of doped fiber into a regular (undoped) optical fiber line. The doped fiber is optically pumped with a second laser wavelength that is coupled into the line in addition to the signal wave. Both wavelengths of light are transmitted through the doped fiber, which transfers energy from the second pump wavelength to the signal wave. The process that causes the amplification is stimulated emission.

Optical fiber is also widely exploited as a nonlinear medium. The glass medium supports a host of nonlinear optical interactions, and the long interaction lengths possible in fiber facilitate a variety of phenomena, which are harnessed for applications and fundamental investigation. Conversely, fiber nonlinearity can have deleterious effects on optical signals, and measures are often required to minimize such unwanted effects.

Optical fibers doped with a wavelength shifter collect scintillation light in physics experiments.

Fiber optic sights for handguns, rifles, and shotguns use pieces of optical fiber to improve visibility of markings on the sight.

Principle of Operation

An overview of the operating principles of the optical fiber

An optical fiber is a cylindrical dielectric waveguide (nonconducting waveguide) that transmits light along its axis, by the process of total internal reflection. The fiber consists of a *core* surrounded by a cladding layer, both of which are made of dielectric materials. To confine the optical signal in the core, the refractive index of the core must be greater than that of the cladding. The boundary between the core and cladding may either be abrupt, in *step-index fiber*, or gradual, in *graded-index fiber*.

Index of Refraction

The index of refraction (or refractive index) is a way of measuring the speed of light in a material. Light travels fastest in a vacuum, such as in outer space. The speed of light in a vacuum is about 300,000 kilometers (186,000 miles) per second. The refractive index of a medium is calculated by dividing the speed of light in a vacuum by the speed of light in that medium. The refractive index of a vacuum is therefore 1, by definition. A typical singlemode fiber used for telecommunications has a cladding made of pure silica, with an index of 1.444 at 1,500 nm, and a core of doped silica with an index around 1.4475. The larger the index of refraction, the slower light travels in that medium. From this information, a simple rule of thumb is that a signal using optical fiber for communication will travel at around 200,000 kilometers per second. To put it another way, the signal will take 5 milliseconds to travel 1,000 kilometers in fiber. Thus a phone call carried by fiber between Sydney and New York, a 16,000-kilometer distance, means that there is a minimum delay of 80 milliseconds (about of a second) between when one caller speaks and the other hears. (The fiber in this case will probably travel a longer route, and there will be additional delays due to communication equipment switching and the process of encoding and decoding the voice onto the fiber).

Total Internal Reflection

When light traveling in an optically dense medium hits a boundary at a steep angle (larger than the critical angle for the boundary), the light is completely reflected. This is called total internal reflection. This effect is used in optical fibers to confine light in the core. Light travels through the fiber core, bouncing back and forth off the boundary between the core and cladding. Because the light must strike the boundary with an angle greater than the critical angle, only light that enters the fiber within a certain range of angles can travel down the fiber without leaking out. This range of angles is called the acceptance cone of the fiber. The size of this acceptance cone is a function of the refractive index difference between the fiber's core and cladding.

In simpler terms, there is a maximum angle from the fiber axis at which light may enter the fiber so that it will propagate, or travel, in the core of the fiber. The sine of this maximum angle is the numerical aperture (NA) of the fiber. Fiber with a larger NA requires less precision to splice and work with than fiber with a smaller NA. Single-mode fiber has a small NA.

Multi-mode Fiber

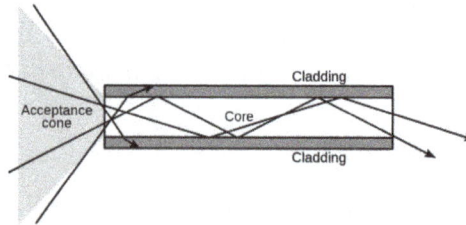

The propagation of light through a multi-mode optical fiber.

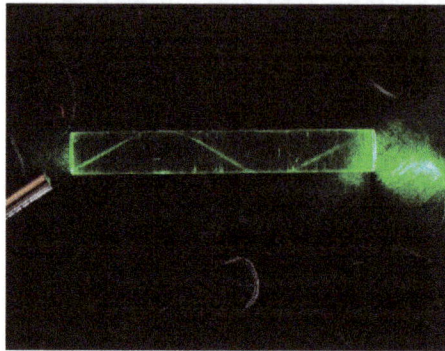

A laser bouncing down an acrylic rod, illustrating the total internal reflection of light in a multi-mode optical fiber.

Fiber with large core diameter (greater than 10 micrometers) may be analyzed by geometrical optics. Such fiber is called *multi-mode fiber*, from the electro-magnetic analysis. In a step-index multi-mode fiber, rays of light are guided along the fiber core by total internal reflection. Rays that meet the core-cladding boundary at a high angle (measured relative to a line normal to the boundary), greater than the critical angle for this boundary, are completely reflected. The critical angle (minimum angle for total internal reflection) is determined by the difference in index of refraction between the core and cladding materials. Rays that meet the boundary at a low angle are refracted from the core into the cladding, and do not convey light and hence information along the fiber. The critical angle determines the acceptance angle of the fiber, often reported as a numerical aperture. A high numerical aperture allows light to propagate down the fiber in rays both close to the axis and at various angles, allowing efficient coupling of light into the fiber. However, this high numerical aperture increases the amount of dispersion as rays at different angles have different path lengths and therefore take different times to traverse the fiber.

In graded-index fiber, the index of refraction in the core decreases continuously be-tween the axis and the cladding. This causes light rays to bend smoothly as they ap-proach the cladding, rather than reflecting abruptly from the core-cladding boundary. The resulting curved paths reduce multi-path dispersion because high angle rays pass more through the lower-index periphery of the core, rather than the high-index center.

The index profile is chosen to minimize the difference in axial propagation speeds of the various rays in the fiber. This ideal index profile is very close to a parabolic relationship between the index and the distance from the axis.

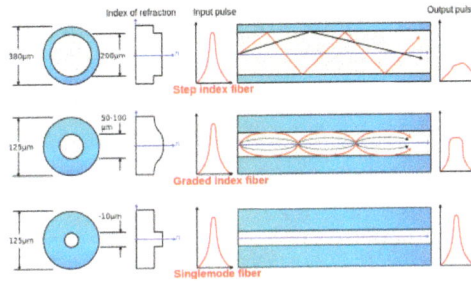

Optical fiber types.

Single-mode Fiber

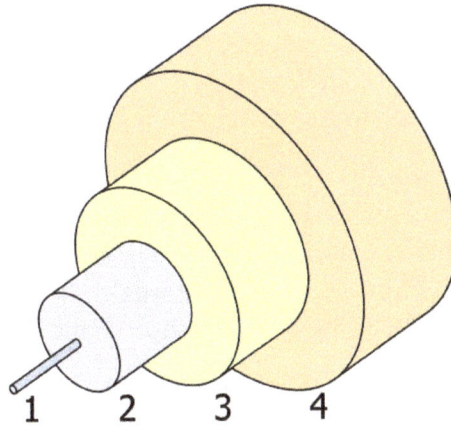

The structure of a typical single-mode fiber.
1. Core: 8 μm diameter
2. Cladding: 125 μm dia.
3. Buffer: 250 μm dia.
4. Jacket: 400 μm dia.

Fiber with a core diameter less than about ten times the wavelength of the propagating light cannot be modeled using geometric optics. Instead, it must be analyzed as an electromagnetic structure, by solution of Maxwell's equations as reduced to the electromagnetic wave equation. The electromagnetic analysis may also be required to understand behaviors such as speckle that occur when coherent light propagates in multi-mode fiber. As an optical waveguide, the fiber supports one or more confined transverse modes by which light can propagate along the fiber. Fiber supporting only one mode is called *single-mode* or *mono-mode fiber*. The behavior of larger-core multi-mode fiber can also be modeled using the wave equation, which shows that such fiber supports more than one mode of propagation (hence the name). The results of such modeling of multi-mode fiber approximately agree with the predictions of geometric optics, if the fiber core is large enough to support more than a few modes.

The waveguide analysis shows that the light energy in the fiber is not completely confined in the core. Instead, especially in single-mode fibers, a significant fraction of the energy in the bound mode travels in the cladding as an evanescent wave.

The most common type of single-mode fiber has a core diameter of 8–10 micrometers and is designed for use in the near infrared. The mode structure depends on the wavelength of the light used, so that this fiber actually supports a small number of additional modes at visible wavelengths. Multi-mode fiber, by comparison, is manufactured with core diameters as small as 50 micrometers and as large as hundreds of micrometers. The normalized frequency V for this fiber should be less than the first zero of the Bessel function J_0 (approximately 2.405).

Special-purpose Fiber

Some special-purpose optical fiber is constructed with a non-cylindrical core and/or cladding layer, usually with an elliptical or rectangular cross-section. These include polarization-maintaining fiber and fiber designed to suppress whispering gallery mode propagation. Polarization-maintaining fiber is a unique type of fiber that is commonly used in fiber optic sensors due to its ability to maintain the polarization of the light inserted into it.

Photonic-crystal fiber is made with a regular pattern of index variation (often in the form of cylindrical holes that run along the length of the fiber). Such fiber uses diffraction effects instead of or in addition to total internal reflection, to confine light to the fiber's core. The properties of the fiber can be tailored to a wide variety of applications.

Mechanisms of Attenuation

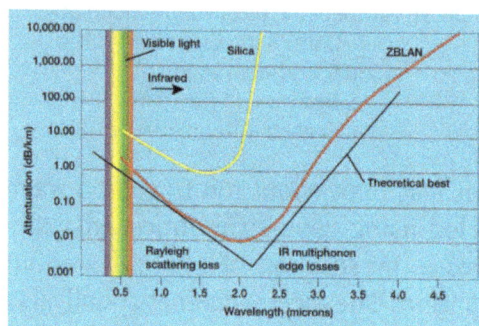

Light attenuation by ZBLAN and silica fibers

Attenuation in fiber optics, also known as transmission loss, is the reduction in intensity of the light beam (or signal) as it travels through the transmission medium. Attenuation coefficients in fiber optics usually use units of dB/km through the medium due to the relatively high quality of transparency of modern optical transmission media. The medium is usually a fiber of silica glass that confines the incident light beam to the inside. Attenuation is an important factor limiting the transmission of a digital signal across large distances.

Thus, much research has gone into both limiting the attenuation and maximizing the amplification of the optical signal. Empirical research has shown that attenuation in optical fiber is caused primarily by both scattering and absorption. Single-mode optical fibers can be made with extremely low loss. Corning's SMF-28 fiber, a standard single-mode fiber for telecommunications wavelengths, has a loss of 0.17 dB/km at 1550 nm. For example, an 8 km length of SMF-28 transmits nearly 75% of light at 1,550 nm. It has been noted that if ocean water was as clear as fiber, one could see all the way to the bottom even of the Marianas Trench in the Pacific Ocean, a depth of 36,000 feet.

Light Scattering

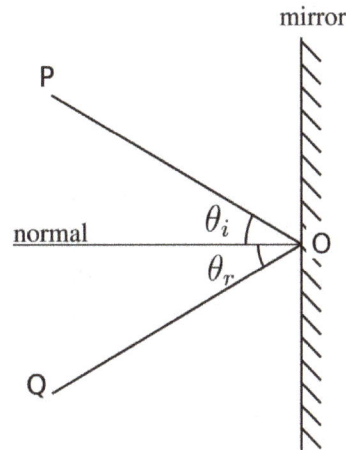

Specular reflection

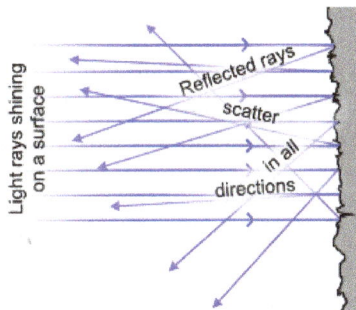

Diffuse reflection

The propagation of light through the core of an optical fiber is based on total internal reflection of the lightwave. Rough and irregular surfaces, even at the molecular level, can cause light rays to be reflected in random directions. This is called diffuse reflection or scattering, and it is typically characterized by wide variety of reflection angles.

Light scattering depends on the wavelength of the light being scattered. Thus, limits to spatial scales of visibility arise, depending on the frequency of the incident light-wave and the physical dimension (or spatial scale) of the scattering center, which is typically in the form of some specific micro-structural feature. Since visible light has a wavelength of the order of one micrometer (one millionth of a meter) scattering centers will have dimensions on a similar spatial scale.

Thus, attenuation results from the incoherent scattering of light at internal surfaces and interfaces. In (poly)crystalline materials such as metals and ceramics, in addition to pores, most of the internal surfaces or interfaces are in the form of grain boundaries that separate tiny regions of crystalline order. It has recently been shown that when the size of the scattering center (or grain boundary) is reduced below the size of the wavelength of the light being scattered, the scattering no longer occurs to any significant extent. This phenomenon has given rise to the production of transparent ceramic materials.

Similarly, the scattering of light in optical quality glass fiber is caused by molecular level irregularities (compositional fluctuations) in the glass structure. Indeed, one emerging school of thought is that a glass is simply the limiting case of a polycrystalline solid. Within this framework, "domains" exhibiting various degrees of short-range order become the building blocks of both metals and alloys, as well as glasses and ceramics. Distributed both between and within these domains are micro-structural defects that provide the most ideal locations for light scattering. This same phenomenon is seen as one of the limiting factors in the transparency of IR missile domes.

At high optical powers, scattering can also be caused by nonlinear optical processes in the fiber.

UV-Vis-IR Absorption

In addition to light scattering, attenuation or signal loss can also occur due to selective absorption of specific wavelengths, in a manner similar to that responsible for the appearance of color. Primary material considerations include both electrons and molecules as follows:

- At the electronic level, it depends on whether the electron orbitals are spaced (or "quantized") such that they can absorb a quantum of light (or photon) of a specific wavelength or frequency in the ultraviolet (UV) or visible ranges. This is what gives rise to color.

- At the atomic or molecular level, it depends on the frequencies of atomic or molecular vibrations or chemical bonds, how close-packed its atoms or molecules are, and whether or not the atoms or molecules exhibit long-range order. These factors will determine the capacity of the material transmitting longer wavelengths in the infrared (IR), far IR, radio and microwave ranges.

The design of any optically transparent device requires the selection of materials based upon knowledge of its properties and limitations. The Lattice absorption characteristics observed at the lower frequency regions (mid IR to far-infrared wavelength range) define the long-wavelength transparency limit of the material. They are the result of the interactive coupling between the motions of thermally induced vibrations of the constituent atoms and molecules of the solid lattice and the incident light wave radiation. Hence, all materials are bounded by limiting regions of absorption caused by atomic and molecular vibrations (bond-stretching)in the far-infrared (>10 μm).

Thus, multi-phonon absorption occurs when two or more phonons simultaneously interact to produce electric dipole moments with which the incident radiation may couple. These dipoles can absorb energy from the incident radiation, reaching a maximum coupling with the radiation when the frequency is equal to the fundamental vibrational mode of the molecular dipole (e.g. Si-O bond) in the far-infrared, or one of its harmonics.

The selective absorption of infrared (IR) light by a particular material occurs because the selected frequency of the light wave matches the frequency (or an integer multiple of the frequency) at which the particles of that material vibrate. Since different atoms and molecules have different natural frequencies of vibration, they will selectively absorb different frequencies (or portions of the spectrum) of infrared (IR) light.

Reflection and transmission of light waves occur because the frequencies of the light waves do not match the natural resonant frequencies of vibration of the objects. When IR light of these frequencies strikes an object, the energy is either reflected or transmitted.

Manufacturing

Materials

Glass optical fibers are almost always made from silica, but some other materials, such as fluorozirconate, fluoroaluminate, and chalcogenide glasses as well as crystalline materials like sapphire, are used for longer-wavelength infrared or other specialized applications. Silica and fluoride glasses usually have refractive indices of about 1.5, but some materials such as the chalcogenides can have indices as high as 3. Typically the index difference between core and cladding is less than one percent.

Plastic optical fibers (POF) are commonly step-index multi-mode fibers with a core diameter of 0.5 millimeters or larger. POF typically have higher attenuation coefficients than glass fibers, 1 dB/m or higher, and this high attenuation limits the range of POF-based systems.

Silica

Silica exhibits fairly good optical transmission over a wide range of wavelengths. In the near-infrared (near IR) portion of the spectrum, particularly around 1.5 μm, silica can have extremely low absorption and scattering losses of the order of 0.2 dB/km. Such remarkably low losses are possible only because ultra-pure silicon is available, it being essential for manufacturing integrated circuits and discrete transistors. A high transparency in the 1.4-µm region is achieved by maintaining a low concentration of hydroxyl groups (OH). Alternatively, a high OH concentration is better for transmission in the ultraviolet (UV) region.

Silica can be drawn into fibers at reasonably high temperatures, and has a fairly broad glass transformation range. One other advantage is that fusion splicing and cleaving of silica fibers is relatively effective. Silica fiber also has high mechanical strength against both pulling and even bending, provided that the fiber is not too thick and that the surfaces have been well prepared during processing. Even simple cleaving (breaking) of the ends of the fiber can provide nicely flat surfaces with acceptable optical quality. Silica is also relatively chemically inert. In particular, it is not hygroscopic (does not absorb water).

Silica glass can be doped with various materials. One purpose of doping is to raise the refractive index (e.g. with germanium dioxide (GeO_2) or aluminium oxide (Al_2O_3)) or to lower it (e.g. with fluorine or boron trioxide (B_2O_3)). Doping is also possible with laser-active ions (for example, rare earth-doped fibers) in order to obtain active fibers to be used, for example, in fiber amplifiers or laser applications. Both the fiber core and cladding are typically doped, so that the entire assembly (core and cladding) is effectively the same compound (e.g. an aluminosilicate, germanosilicate, phosphosilicate or borosilicate glass).

Particularly for active fibers, pure silica is usually not a very suitable host glass, because it exhibits a low solubility for rare earth ions. This can lead to quenching effects due to clustering of dopant ions. Aluminosilicates are much more effective in this respect.

Silica fiber also exhibits a high threshold for optical damage. This property ensures a low tendency for laser-induced breakdown. This is important for fiber amplifiers when utilized for the amplification of short pulses.

Because of these properties silica fibers are the material of choice in many optical applications, such as communications (except for very short distances with plastic optical fiber), fiber lasers, fiber amplifiers, and fiber-optic sensors. Large efforts put forth in the development of various types of silica fibers have further increased the performance of such fibers over other materials.

Fluoride Glass

Fluoride glass is a class of non-oxide optical quality glasses composed of fluorides of various metals. Because of their low viscosity, it is very difficult to completely avoid crystallization while processing it through the glass transition (or drawing the fiber from the melt). Thus, although heavy metal fluoride glasses (HMFG) exhibit very low optical attenuation, they are not only difficult to manufacture, but are quite fragile, and have poor resistance to moisture and other environmental attacks. Their best attribute is that they lack the absorption band associated with the hydroxyl (OH) group (3,200–3,600 cm^{-1}; i.e., 2,777–3,125 nm or 2.78–3.13 µm), which is present in nearly all oxide-based glasses.

An example of a heavy metal fluoride glass is the ZBLAN glass group, composed of zirconium, barium, lanthanum, aluminium, and sodium fluorides. Their main technological application is as optical waveguides in both planar and fiber form. They are advantageous especially in the mid-infrared (2,000–5,000 nm) range.

HMFGs were initially slated for optical fiber applications, because the intrinsic losses of a mid-IR fiber could in principle be lower than those of silica fibers, which are transparent only up to about 2 µm. However, such low losses were never realized in practice, and the fragility and high cost of fluoride fibers made them less than ideal as primary candidates. Later, the utility of fluoride fibers for various other applications was discovered. These include mid-IR spectroscopy, fiber optic sensors, thermometry, and imaging. Also, fluoride fibers can be used for guided lightwave transmission in media such as YAG (yttrium aluminium garnet) lasers at 2.9 µm, as required for medical applications (e.g. ophthalmology and dentistry).

Phosphate Glass

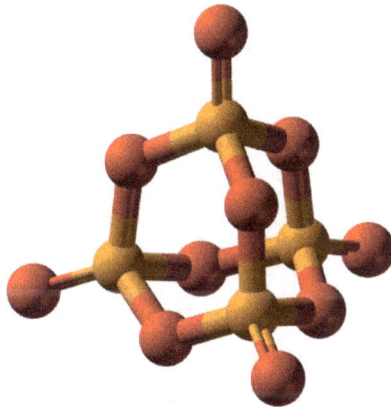

The P4O10 cagelike structure—the basic building block for phosphate glass.

Phosphate glass constitutes a class of optical glasses composed of metaphosphates of various metals. Instead of the SiO_4 tetrahedra observed in silicate glasses, the building block for this glass former is phosphorus pentoxide (P_2O_5), which crystallizes in at least four different forms. The most familiar polymorph comprises molecules of P_4O_{10}.

Phosphate glasses can be advantageous over silica glasses for optical fibers with a high concentration of doping rare earth ions. A mix of fluoride glass and phosphate glass is fluorophosphate glass.

Chalcogenide Glass

The chalcogens—the elements in group 16 of the periodic table—particularly sulfur (S), selenium (Se) and tellurium (Te)—react with more electropositive elements, such as silver, to form chalcogenides. These are extremely versatile compounds, in that they can be crystalline or amorphous, metallic or semiconducting, and conductors of ions or electrons. Glass containing chalcogenides can be used to make fibers for far infrared transmission.

Process

Preform

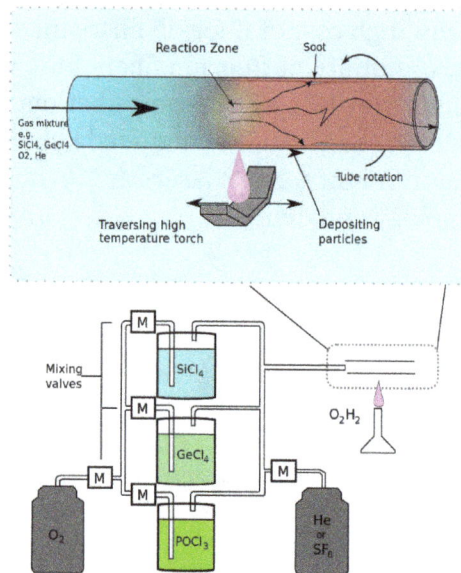

Illustration of the modified chemical vapor deposition (inside) process

Standard optical fibers are made by first constructing a large-diameter "preform" with a carefully controlled refractive index profile, and then "pulling" the preform to form the long, thin optical fiber. The preform is commonly made by three chemical vapor deposition methods: *inside vapor deposition*, *outside vapor deposition*, and *vapor axial deposition*.

With *inside vapor deposition*, the preform starts as a hollow glass tube approximately 40 centimeters (16 in) long, which is placed horizontally and rotated slowly on a lathe. Gases such as silicon tetrachloride ($SiCl_4$) or germanium tetrachloride ($GeCl_4$) are injected with oxygen in the end of the tube. The gases are then heated by means of an external hydrogen burner, bringing the temperature of the gas up to 1,900 K (1,600 °C, 3,000 °F), where the tetrachlorides react with oxygen to produce silica or germania (germanium dioxide) particles. When the reaction conditions are chosen to allow this reaction to occur in the gas phase throughout the tube volume, in contrast to earlier techniques where the reaction occurred only on the glass surface, this technique is called *modified chemical vapor deposition (MCVD)*.

The oxide particles then agglomerate to form large particle chains, which subsequently deposit on the walls of the tube as soot. The deposition is due to the large difference in temperature between the gas core and the wall causing the gas to push the particles outwards (this is known as thermophoresis). The torch is then traversed up and down the length of the tube to deposit the material evenly. After the torch has reached the end of the tube, it is then brought back to the beginning of the tube and the deposited particles are then melted to form a solid layer. This process is repeated until a sufficient amount of material has been deposited. For each layer the composition can be modified by varying the gas composition, resulting in precise control of the finished fiber's optical properties.

In outside vapor deposition or vapor axial deposition, the glass is formed by *flame hydrolysis*, a reaction in which silicon tetrachloride and germanium tetrachloride are oxidized by reaction with water (H_2O) in an oxyhydrogen flame. In outside vapor deposition the glass is deposited onto a solid rod, which is removed before further processing. In vapor axial deposition, a short *seed rod* is used, and a porous preform, whose length is not limited by the size of the source rod, is built up on its end. The porous preform is consolidated into a transparent, solid preform by heating to about 1,800 K (1,500 °C, 2,800 °F).

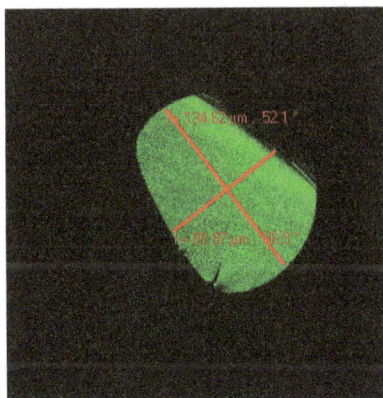

Cross-section of a fiber drawn from a D-shaped preform

Typical communications fiber uses a circular preform. For some applications such as double-clad fibers another form is preferred. In fiber lasers based on double-clad fiber, an asymmetric shape improves the filling factor for laser pumping.

Because of the surface tension, the shape is smoothed during the drawing process, and the shape of the resulting fiber does not reproduce the sharp edges of the preform. Nevertheless, careful polishing of the preform is important, since any defects of the preform surface affect the optical and mechanical properties of the resulting fiber. In particular, the preform for the test-fiber shown in the figure was not polished well, and cracks are seen with the confocal optical microscope.

Drawing

The preform, however constructed, is placed in a device known as a drawing tower, where the preform tip is heated and the optical fiber is pulled out as a string. By measuring the resultant fiber width, the tension on the fiber can be controlled to maintain the fiber thickness.

Coatings

The light is guided down the core of the fiber by an optical cladding with a lower refractive index that traps light in the core through total internal reflection.

The cladding is coated by a buffer that protects it from moisture and physical damage. The buffer coating is what gets stripped off the fiber for termination or splicing. These coatings are UV-cured urethane acrylate composite or polyimide materials applied to the outside of the fiber during the drawing process. The coatings protect the very delicate strands of glass fiber—about the size of a human hair—and allow it to survive the rigors of manufacturing, proof testing, cabling and installation.

Today's glass optical fiber draw processes employ a dual-layer coating approach. An inner primary coating is designed to act as a shock absorber to minimize attenuation caused by microbending. An outer secondary coating protects the primary coating against mechanical damage and acts as a barrier to lateral forces. Sometimes a metallic armor layer is added to provide extra protection.

These fiber optic coating layers are applied during the fiber draw, at speeds approaching 100 kilometers per hour (60 mph). Fiber optic coatings are applied using one of two methods: *wet-on-dry* and *wet-on-wet*. In wet-on-dry, the fiber passes through a primary coating application, which is then UV cured—then through the secondary coating application, which is subsequently cured. In wet-on-wet, the fiber passes through both the primary and secondary coating applications, then goes to UV curing.

Fiber optic coatings are applied in concentric layers to prevent damage to the fiber during the drawing application and to maximize fiber strength and microbend resistance. Unevenly coated fiber will experience non-uniform forces when the coating expands or contracts, and is susceptible to greater signal attenuation. Under proper drawing and coating processes, the coatings are concentric around the fiber, continuous over the length of the application and have constant thickness.

Fiber optic coatings protect the glass fibers from scratches that could lead to strength degradation. The combination of moisture and scratches accelerates the aging and deterioration of fiber strength. When fiber is subjected to low stresses over a long period, fiber fatigue can occur. Over time or in extreme conditions, these factors combine to cause microscopic flaws in the glass fiber to propagate, which can ultimately result in fiber failure.

Three key characteristics of fiber optic waveguides can be affected by environmental conditions: strength, attenuation and resistance to losses caused by microbending. External fiber optic coatings protect glass optical fiber from environmental conditions that can affect the fiber's performance and long-term durability. On the inside, coatings ensure the reliability of the signal being carried and help minimize attenuation due to microbending.

Practical Issues

Cable Construction

An optical fiber cable

In practical fibers, the cladding is usually coated with a tough resin coating and an additional *buffer* layer, which may be further surrounded by a *jacket* layer, usually plastic. These layers add strength to the fiber but do not contribute to its optical wave guide properties. Rigid fiber assemblies sometimes put light-absorbing ("dark") glass between the fibers, to prevent light that leaks out of one fiber from entering another. This reduces cross-talk between the fibers, or reduces flare in fiber bundle imaging applications.

Modern cables come in a wide variety of sheathings and armor, designed for applications such as direct burial in trenches, high voltage isolation, dual use as power lines,installation in conduit, lashing to aerial telephone poles, submarine installation, and insertion in paved streets. Multi-fiber cable usually uses colored coatings and/or buffers to identify each strand. The cost of small fiber-count pole-mounted cables has greatly decreased due to the high demand for fiber to the home (FTTH) installations in Japan and South Korea.

Fiber cable can be very flexible, but traditional fiber's loss increases greatly if the fiber is bent with a radius smaller than around 30 mm. This creates a problem when the cable is bent around corners or wound around a spool, making FTTX installations more complicated. "Bendable fibers", targeted towards easier installation in home environments, have been standardized as ITU-T G.657. This type of fiber can be bent with a radius as low as 7.5 mm without adverse impact. Even more bendable fibers have been developed. Bendable fiber may also be resistant to fiber hacking, in which the signal in a fiber is surreptitiously monitored by bending the fiber and detecting the leakage.

Another important feature of cable is cable's ability to withstand horizontally applied force. It is technically called max tensile strength defining how much force can be applied to the cable during the installation period.

Some fiber optic cable versions are reinforced with aramid yarns or glass yarns as intermediary strength member. In commercial terms, usage of the glass yarns are more cost effective while no loss in mechanical durability of the cable. Glass yarns also protect the cable core against rodents and termites.

Termination and Splicing

ST connectors on multi-mode fiber.

Optical fibers are connected to terminal equipment by optical fiber connectors. These connectors are usually of a standard type such as *FC*, *SC*, *ST*, *LC*, *MTRJ*, or *SMA*. Optical fibers may be connected to each other by connectors or by *splicing*, that is, joining two fibers together to form a continuous optical waveguide. The generally accepted splicing method is arc fusion splicing, which melts the fiber ends together with an electric arc. For quicker fastening jobs, a "mechanical splice" is used.

Fusion splicing is done with a specialized instrument. The fiber ends are first stripped of their protective polymer coating (as well as the more sturdy outer jacket, if present). The ends are *cleaved* (cut) with a precision cleaver to make them perpendicular, and are placed into special holders in the fusion splicer. The splice is usually inspected via a magnified viewing screen to check the cleaves before and after the splice. The splicer uses small motors to align the end faces together, and emits a small spark between electrodes at the gap to burn off dust and moisture. Then the splicer generates a larger spark that raises the temperature above the melting point of the glass, fusing the ends together permanently. The location and energy of the spark is carefully controlled so that the molten core and cladding do not mix, and this minimizes optical loss. A splice loss estimate is measured by the splicer, by directing light through the cladding on one side and measuring the light leaking from the cladding on the other side. A splice loss under 0.1 dB is typical. The complexity of this process makes fiber splicing much more difficult than splicing copper wire.

Mechanical fiber splices are designed to be quicker and easier to install, but there is still the need for stripping, careful cleaning and precision cleaving. The fiber ends are aligned and held together by a precision-made sleeve, often using a clear index-matching gel that enhances the transmission of light across the joint. Such joints typically have higher optical loss and are less robust than fusion splices, especially if the gel is used. All splicing techniques involve installing an enclosure that protects the splice.

Fibers are terminated in connectors that hold the fiber end precisely and securely. A fiber-optic connector is basically a rigid cylindrical barrel surrounded by a sleeve that holds the barrel in its mating socket. The mating mechanism can be *push and click*, *turn and latch* (*bayonet mount*), or *screw-in* (*threaded*). The barrel is typically free to move within the sleeve, and may have a key that prevents the barrel and fiber from rotating as the connectors are mated.

A typical connector is installed by preparing the fiber end and inserting it into the rear of the connector body. Quick-set adhesive is usually used to hold the fiber securely, and a strain relief is secured to the rear. Once the adhesive sets, the fiber's end is polished to a mirror finish. Various polish profiles are used, depending on the type of fiber and the application. For single-mode fiber, fiber ends are typically polished with a slight curvature that makes the mated connectors touch only at their cores. This is called a *physical contact* (PC) polish. The curved surface may be polished at an angle, to make an *angled physical contact (APC)* connection. Such connections have higher loss than PC connections, but greatly reduced back reflection, because light that reflects from the angled surface leaks out of the fiber core. The resulting signal strength loss is called *gap loss*. APC fiber ends have low back reflection even when disconnected.

In the 1990s, terminating fiber optic cables was labor-intensive. The number of parts per connector, polishing of the fibers, and the need to oven-bake the epoxy in each connector made terminating fiber optic cables difficult. Today, many connectors types are on the market that offer easier, less labor-intensive ways of terminating cables. Some of the most popular connectors are pre-polished at the factory, and include a gel inside the connector. Those two steps help save money on labor, especially on large projects. A cleave is made at a required length, to get as close to the polished piece already inside the connector. The gel surrounds the point where the two pieces meet inside the connector for very little light loss.

Free-space Coupling

It is often necessary to align an optical fiber with another optical fiber, or with an optoelectronic device such as a light-emitting diode, a laser diode, or a modulator. This can involve either carefully aligning the fiber and placing it in contact with the device, or can use a lens to allow coupling over an air gap. Typically the size of the fiber mode is much larger than the size of the mode in a laser diode or a silicon optical chip. In this case a tapered or lensed fiber is used to match the fiber mode field distribution to that of the other element. The lens on the end of the fiber can be formed using polishing, laser cutting or fusion splicing.

In a laboratory environment, a bare fiber end is coupled using a fiber launch system, which uses a microscope objective lens to focus the light down to a fine point. A precision translation stage (micro-positioning table) is used to move the lens, fiber, or device to allow the coupling efficiency to be optimized. Fibers with a connector on the end make this process much simpler: the connector is simply plugged into a pre-aligned fiberoptic collimator, which contains a lens that is either accurately positioned with respect to the fiber, or is adjustable. To achieve the best injection efficiency into single-mode fiber, the direction, position, size and divergence of the beam must all be optimized. With good beams, 70 to 90% coupling efficiency can be achieved.

With properly polished single-mode fibers, the emitted beam has an almost perfect Gaussian shape—even in the far field—if a good lens is used. The lens needs to be large enough to support the full numerical aperture of the fiber, and must not introduce aberrations in the beam. Aspheric lenses are typically used.

Fiber Fuse

At high optical intensities, above 2 megawatts per square centimeter, when a fiber is subjected to a shock or is otherwise suddenly damaged, a *fiber fuse* can occur. The reflection from the damage vaporizes the fiber immediately before the break, and this new defect remains reflective so that the damage propagates back toward the transmitter at 1–3 meters per second (4–11 km/h, 2–8 mph). The open fiber control system, which ensures laser eye safety in the event of a broken fiber, can also effectively halt propagation of the fiber fuse. In situations, such as undersea cables, where high power levels might be used without the need for open fiber control, a "fiber fuse" protection device at the transmitter can break the circuit to keep damage to a minimum.

Chromatic Dispersion

The refractive index of fibers varies slightly with the frequency of light, and light sources are not perfectly monochromatic. Modulation of the light source to transmit a signal also slightly widens the frequency band of the transmitted light. This has the effect that, over long distances and at high modulation speeds, the different frequencies of light can take different times to arrive at the receiver, ultimately making the signal impossible to discern, and requiring extra repeaters. This problem can be overcome in a number of ways, including the use of a relatively short length of fiber that has the opposite refractive index gradient.

Fiber-optic Communication

Fiber-optic communication is a method of transmitting information from one place to another by sending pulses of light through an optical fiber. The light forms an electromagnetic carrier wave that is modulated to carry information. First developed in the 1970s, fiber-optics have revolutionized the telecommunications industry and have played a major role in the advent of the Information Age. Because of its advantages over electrical transmission, optical fibers have largely replaced copper wire communications in core networks in the developed world. Optical fiber is used by many telecommunications companies to transmit telephone signals, Internet communication, and cable television signals. Researchers at Bell Labs have reached internet speeds of over 100 petabit×kilometer per second using fiber-optic communication.

The process of communicating using fiber-optics involves the following basic steps:

1. creating the optical signal involving the use of a transmitter, usually from an electrical signal

2. relaying the signal along the fiber, ensuring that the signal does not become too distorted or weak

3. receiving the optical signal

4. converting it into an electrical signal

Applications

Optical fiber is used by many telecommunications companies to transmit telephone signals, Internet communication, and cable television signals. Due to much lower attenuation and interference, optical fiber has large advantages over existing copper wire in long-distance and high-demand applications. However, infrastructure development within cities was relatively difficult and time-consuming, and fiber-optic systems were complex and expensive to install and operate. Due to these difficulties, fiber-optic communication systems have primarily been installed in long-distance applications, where they can be used to their full transmission capacity, offsetting the increased cost. Since 2000, the prices for fiber-optic communications have dropped considerably.

The price for rolling out fiber to the home has currently become more cost-effective than that of rolling out a copper based network. Prices have dropped to $850 per subscriber in the US and lower in countries like The Netherlands, where digging costs are low and housing density is high.

Since 1990, when optical-amplification systems became commercially available, the telecommunications industry has laid a vast network of intercity and transoceanic fiber communication lines. By 2002, an intercontinental network of 250,000 km of submarine communications cable with a capacity of 2.56 Tb/s was completed, and although specific network capacities are privileged information, telecommunications investment reports indicate that network capacity has increased dramatically since 2004.

History

In 1880 Alexander Graham Bell and his assistant Charles Sumner Tainter created a very early precursor to fiber-optic communications, the Photophone, at Bell's newly established Volta Laboratory in Washington, D.C. Bell considered it his most important invention. The device allowed for the transmission of sound on a beam of light. On June 3, 1880, Bell conducted the world's first wireless telephone transmission between two buildings, some 213 meters apart. Due to its use of an atmospheric transmission medium, the Photophone would not prove practical until advances in laser and optical fiber technologies permitted the secure transport of light. The Photophone's first practical use came in military communication systems many decades later.

In 1954 Harold Hopkins and Narinder Singh Kapany showed that rolled fiber glass allowed light to be transmitted. Initially it was considered that the light can traverse in only straight medium.

In 1966 Charles K. Kao and George Hockham proposed optical fibers at STC Laboratories (STL) at Harlow, England, when they showed that the losses of 1,000 dB/km in existing glass (compared to 5-10 dB/km in coaxial cable) were due to contaminants, which could potentially be removed.

Optical fiber was successfully developed in 1970 by Corning Glass Works, with attenuation low enough for communication purposes (about 20 dB/km), and at the same time GaAs semiconductor lasers were developed that were compact and therefore suitable for transmitting light through fiber optic cables for long distances.

After a period of research starting from 1975, the first commercial fiber-optic communications system was developed, which operated at a wavelength around 0.8 µm and used GaAs semiconductor lasers. This first-generation system operated at a bit rate of 45 Mbps with repeater spacing of up to 10 km. Soon on 22 April 1977, General Telephone and Electronics sent the first live telephone traffic through fiber optics at a 6 Mbit/s throughput in Long Beach, California.

In October 1973, Corning Glass signed a development contract with CSELT and Pirelli aimed to test fiber optics in an urban environment: in September 1977, the second cable in this test series, named COS-2, was experimentally deployed in two lines (9 km) in Turin, for the first time in a big city, at a speed of 140 Mbit/s.

The second generation of fiber-optic communication was developed for commercial use in the early 1980s, operated at 1.3 µm, and used InGaAsP semiconductor lasers. These early systems were initially limited by multi mode fiber dispersion, and in 1981 the single-mode fiber was revealed to greatly improve system performance, however practical connectors capable of working with single mode fiber proved difficult to develop. In 1984, they had already developed a fiber optic cable that would help further their progress toward making fiber optic cables that would circle the globe. Canadian service provider SaskTel had completed construction of what was then the world's longest commercial fiberoptic network, which covered 3,268 km and linked 52 communities. By 1987, these systems were operating at bit rates of up to 1.7 Gb/s with repeater spacing up to 50 km.

The first transatlantic telephone cable to use optical fiber was TAT-8, based on Desurvire optimised laser amplification technology. It went into operation in 1988.

Third-generation fiber-optic systems operated at 1.55 µm and had losses of about 0.2 dB/km. This development was spurred by the discovery of Indium gallium arsenide and the development of the Indium Gallium Arsenide photodiode by Pearsall. Engineers overcame earlier difficulties with pulse-spreading at that wavelength using conventional InGaAsP semiconductor lasers. Scientists overcame this difficulty by using dispersion-shifted fibers designed to have minimal dispersion at 1.55 µm or by limiting the laser spectrum to a single longitudinal mode. These developments eventually allowed third-generation systems to operate commercially at 2.5 Gbit/s with repeater spacing in excess of 100 km.

The fourth generation of fiber-optic communication systems used optical amplification to reduce the need for repeaters and wavelength-division multiplexing to increase data capacity. These two improvements caused a revolution that resulted in the doubling of system capacity every six months starting in 1992 until a bit rate of 10 Tb/s was reached by 2001. In 2006 a bit-rate of 14 Tbit/s was reached over a single 160 km line using optical amplifiers.

The focus of development for the fifth generation of fiber-optic communications is on extending the wavelength range over which a WDM system can operate. The conventional wavelength window, known as the C band, covers the wavelength range 1.53-1.57 µm, and *dry fiber* has a low-loss window promising an extension of that range to 1.30-1.65 µm. Other developments include the concept of "optical solitons", pulses that preserve their shape by counteracting the effects of dispersion with the nonlinear effects of the fiber by using pulses of a specific shape.

In the late 1990s through 2000, industry promoters, and research companies such as KMI, and RHK predicted massive increases in demand for communications bandwidth due to increased use of the Internet, and commercialization of various bandwidth-intensive consumer services, such as video on demand. Internet protocol data traffic was increasing exponentially, at a faster rate than integrated circuit complexity had increased under Moore's Law. From the bust of the dot-com bubble through 2006, however, the main trend in the industry has been consolidation of firms and offshoring of manufacturing to reduce costs. Companies such as Verizon and AT&T have taken advantage of fiber-optic communications to deliver a variety of high-throughput data and broadband services to consumers' homes.

Technology

Modern fiber-optic communication systems generally include an optical transmitter to convert an electrical signal into an optical signal to send into the optical fiber, a cable containing bundles of multiple optical fibers that is routed through underground conduits and buildings, multiple kinds of amplifiers, and an optical receiver to recover the signal as an electrical signal. The information transmitted is typically digital information generated by computers, telephone systems, and cable television companies.

Transmitters

A GBIC module (shown here with its cover removed), is an optical and electrical transceiver. The electrical connector is at top right, and the optical connectors are at bottom left

The most commonly used optical transmitters are semiconductor devices such as light-emitting diodes (LEDs) and laser diodes. The difference between LEDs and laser diodes is that LEDs produce incoherent light, while laser diodes produce coherent light. For use in optical communications, semiconductor optical transmitters must be designed to be compact, efficient, and reliable, while operating in an optimal wavelength range, and directly modulated at high frequencies.

In its simplest form, an LED is a forward-biased p-n junction, emitting light through spontaneous emission, a phenomenon referred to as electroluminescence. The emitted light is incoherent with a relatively wide spectral width of 30-60 nm. LED light transmission is also inefficient, with only about 1% of input power, or about 100 microwatts, eventually converted into launched power which has been coupled into the optical fiber. However, due to their relatively simple design, LEDs are very useful for low-cost applications.

Communications LEDs are most commonly made from Indium gallium arsenide phosphide (InGaAsP) or gallium arsenide (GaAs). Because InGaAsP LEDs operate at a longer wavelength than GaAs LEDs (1.3 micrometers vs. 0.81-0.87 micrometers), their output spectrum, while equivalent in energy is wider in wavelength terms by a factor of about 1.7. The large spectrum width of LEDs is subject to higher fiber dispersion, considerably limiting their bit rate-distance product (a common measure of usefulness). LEDs are suitable primarily for local-area-network applications with bit rates of 10-100 Mbit/s and transmission distances of a few kilometers. LEDs have also been developed that use several quantum wells to emit light at different wavelengths over a broad spectrum, and are currently in use for local-area WDM (Wavelength-Division Multiplexing) networks.

Today, LEDs have been largely superseded by VCSEL (Vertical Cavity Surface Emitting Laser) devices, which offer improved speed, power and spectral properties, at a similar cost. Common VCSEL devices couple well to multi mode fiber.

A semiconductor laser emits light through stimulated emission rather than spontaneous emission, which results in high output power (~100 mW) as well as other benefits related to the nature of coherent light. The output of a laser is relatively directional, allowing high coupling efficiency (~50 %) into single-mode fiber. The narrow spectral width also allows for high bit rates since it reduces the effect of chromatic dispersion. Furthermore, semiconductor lasers can be modulated directly at high frequencies because of short recombination time.

Commonly used classes of semiconductor laser transmitters used in fiber optics include VCSEL (Vertical-Cavity Surface-Emitting Laser), Fabry–Pérot and DFB (Distributed Feed Back).

Laser diodes are often directly modulated, that is the light output is controlled by a current applied directly to the device. For very high data rates or very long distance *links*, a laser source may be operated continuous wave, and the light modulated by an external device such as an electro-absorption modulator or Mach–Zehnder interferometer. External modulation increases the achievable link distance by eliminating laser chirp, which broadens the linewidth of directly modulated lasers, increasing the chromatic dispersion in the fiber.

A transceiver is a device combining a transmitter and a receiver in a single housing.

Receivers

The main component of an optical receiver is a photodetector, which converts light into electricity using the photoelectric effect. The primary photodetectors for telecommunications are made from Indium gallium arsenide The photodetector is typically a semiconductor-based photodiode. Several types of photodiodes include p-n photodiodes, p-i-n photodiodes, and avalanche photodiodes. Metal-semiconductor-metal (MSM) photodetectors are also used due to their suitability for circuit integration in regenerators and wavelength-division multiplexers.

Optical-electrical converters are typically coupled with a transimpedance amplifier and a limiting amplifier to produce a digital signal in the electrical domain from the incoming optical signal, which may be attenuated and distorted while passing through the channel. Further signal processing such as clock recovery from data (CDR) performed by a phase-locked loop may also be applied before the data is passed on.

Fiber Cable Types

A cable reel trailer with conduit that can carry optical fiber

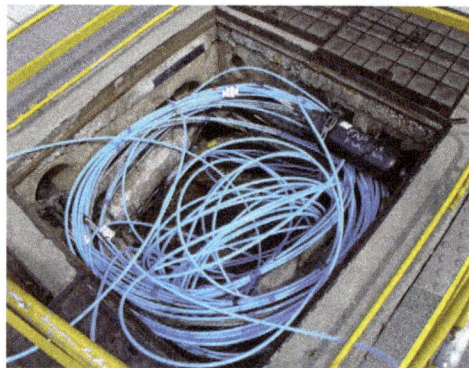

Multi-mode optical fiber in an underground service pit

An optical fiber cable consists of a core, cladding, and a buffer (a protective outer coating), in which the cladding guides the light along the core by using the method of total internal reflection. The core and the cladding (which has a lower-refractive-index) are usually made of high-quality silica glass, although they can both be made of plastic as well. Connecting two optical fibers is done by fusion splicing or mechanical splicing and requires special skills and interconnection technology due to the microscopic precision required to align the fiber cores.

Two main types of optical fiber used in optic communications include multi-mode optical fibers and single-mode optical fibers. A multi-mode optical fiber has a larger core (≥ 50 micrometers), allowing less precise, cheaper transmitters and receivers to connect to it as well as cheaper connectors. However, a multi-mode fiber introduces multimode distortion, which often limits the bandwidth and length of the link. Furthermore, because of its higher dopant content, multi-mode fibers are usually expensive and exhibit higher attenuation. The core of a single-mode fiber is smaller (<10 micrometers) and requires more expensive components and interconnection methods, but allows much longer, higher-performance links.

In order to package fiber into a commercially viable product, it typically is protectively coated by using ultraviolet (UV), light-cured acrylate polymers, then terminated with optical fiber connectors, and finally assembled into a cable. After that, it can be laid in the ground and then run through the walls of a building and deployed aerially in a manner similar to copper cables. These fibers require less maintenance than common twisted pair wires, once they are deployed.

Specialized cables are used for long distance subsea data transmission, e.g. transatlantic communications cable. New (2011–2013) cables operated by commercial enterprises (Emerald Atlantis, Hibernia Atlantic) typically have four strands of fiber and cross the Atlantic (NYC-London) in 60-70ms. Cost of each such cable was about $300M in 2011. *source: The Chronicle Herald.*

Another common practice is to bundle many fiber optic strands within long-distance power transmission cable. This exploits power transmission rights of way effectively, ensures a power company can own and control the fiber required to monitor its own devices and lines, is effectively immune to tampering, and simplifies the deployment of smart grid technology.

Amplifier

The transmission distance of a fiber-optic communication system has traditionally been limited by fiber attenuation and by fiber distortion. By using opto-electronic repeaters, these problems have been eliminated. These repeaters convert the signal into an electrical signal, and then use a transmitter to send the signal again at a higher intensity than was received, thus counteracting the loss incurred in the previous segment. Because of the high complexity with modern wavelength-division multiplexed signals (including the fact that they had to be installed about once every 20 km), the cost of these repeaters is very high.

An alternative approach is to use an optical amplifier, which amplifies the optical signal directly without having to convert the signal into the electrical domain. It is made by doping a length of fiber with the rare-earth mineral erbium, and *pumping* it with light from a laser with a shorter wavelength than the communications signal (typically 980 nm). Amplifiers have largely replaced repeaters in new installations.

Wavelength-division Multiplexing

Wavelength-division multiplexing (WDM) is the practice of multiplying the available capacity of optical fibers through use of parallel channels, each channel on a dedicated wavelength of light. This requires a wavelength division multiplexer in the transmitting equipment and a demultiplexer (essentially a spectrometer) in the receiving equipment. Arrayed waveguide gratings are commonly used for multiplexing and demultiplexing in WDM. Using WDM technology now commercially available, the bandwidth of a fiber can be divided into as many as 160 channels to support a combined bit rate in the range of 1.6 Tbit/s.

Parameters

Bandwidth–distance Product

Because the effect of dispersion increases with the length of the fiber, a fiber transmission system is often characterized by its *bandwidth–distance product*, usually expressed in units of MHz·km. This value is a product of bandwidth and distance because there is a trade off between the bandwidth of the signal and the distance it can be carried. For example, a common multi-mode fiber with bandwidth–distance product of 500 MHz·km could carry a 500 MHz signal for 1 km or a 1000 MHz signal for 0.5 km.

Engineers are always looking at current limitations in order to improve fiber-optic communication, and several of these restrictions are currently being researched.

Record Speeds

Each fiber can carry many independent channels, each using a different wavelength of light (wavelength-division multiplexing). The net data rate (data rate without overhead bytes) per fiber is the per-channel data rate reduced by the FEC overhead, multiplied by the number of channels (usually up to eighty in commercial dense WDM systems as of 2008).

Year	Organization	Effective speed	WDM channels	Per channel speed	Distance
2009	Alcatel-Lucent	15 Tbit/s	155	100 Gbit/s	90 km
2010	NTT	69.1 Tbit/s	432	171 Gbit/s	240 km
2011	KIT	26 Tbit/s	1	26 Tbit/s	50 km
2011	NEC	101 Tbit/s	370	273 Gbit/s	165 km
2012	NEC, Corning	1.05 Petabit/s	12 core fiber		52.4 km

While the physical limitations of electrical cable prevent speeds in excess of 10 Gigabits per second, the physical limitations of fiber optics have not yet been reached.In 2013, *New Scientist* reported that a team at the University of Southampton had achieved a throughput of 73.7 Tbit per second, with the signal traveling at 99.7% the speed of light through a hollow-core photonic crystal fiber.

Dispersion

For modern glass optical fiber, the maximum transmission distance is limited not by direct material absorption but by several types of dispersion, or spreading of optical pulses as they travel along the fiber. Dispersion in optical fibers is caused by a variety of factors. Intermodal dispersion, caused by the different axial speeds of different transverse modes, limits the performance of multi-mode fiber. Because single-mode fiber supports only one transverse mode, intermodal dispersion is eliminated.

In single-mode fiber performance is primarily limited by chromatic dispersion (also called group velocity dispersion), which occurs because the index of the glass varies slightly depending on the wavelength of the light, and light from real optical transmitters necessarily has nonzero spectral width (due to modulation). Polarization mode dispersion, another source of limitation, occurs because although the single-mode fiber can sustain only one transverse mode, it can carry this mode with two different polarizations, and slight imperfections or distortions in a fiber can alter the propagation velocities for the two polarizations. This phenomenon is called fiber birefringence and can be counteracted by polarization-maintaining optical fiber. Dispersion limits the bandwidth of the fiber because the spreading optical pulse limits the rate that pulses can follow one another on the fiber and still be distinguishable at the receiver.

Some dispersion, notably chromatic dispersion, can be removed by a 'dispersion compensator'. This works by using a specially prepared length of fiber that has the opposite dispersion to that induced by the transmission fiber, and this sharpens the pulse so that it can be correctly decoded by the electronics.

Attenuation

Fiber attenuation, which necessitates the use of amplification systems, is caused by a combination of material absorption, Rayleigh scattering, Mie scattering, and connection losses. Although material absorption for pure silica is only around 0.03 dB/km (modern fiber has attenuation around 0.3 dB/km), impurities in the original optical fibers caused attenuation of about 1000 dB/km. Other forms of attenuation are caused by physical stresses to the fiber, microscopic fluctuations in density, and imperfect splicing techniques.

Transmission Windows

Each effect that contributes to attenuation and dispersion depends on the optical wavelength. There are wavelength bands (or windows) where these effects are weakest, and these are the most favorable for transmission. These windows have been standardized, and the currently defined bands are the following:

Band	Description	Wavelength Range
O band	original	1260 to 1360 nm
E band	extended	1360 to 1460 nm
S band	short wavelengths	1460 to 1530 nm
C band	conventional ("erbium window")	1530 to 1565 nm
L band	long wavelengths	1565 to 1625 nm
U band	ultralong wavelengths	1625 to 1675 nm

Note that this table shows that current technology has managed to bridge the second and third windows that were originally disjoint.

Historically, there was a window used below the O band, called the first window, at 800-900 nm; however, losses are high in this region so this window is used primarily for short-distance communications. The current lower windows (O and E) around 1300 nm have much lower losses. This region has zero dispersion. The middle windows (S and C) around 1500 nm are the most widely used. This region has the lowest attenuation losses and achieves the longest range. It does have some dispersion, so dispersion compensator devices are used to remove this.

Regeneration

When a communications link must span a larger distance than existing fiber-optic technology is capable of, the signal must be *regenerated* at intermediate points in the link by optical communications repeaters. Repeaters add substantial cost to a communication system, and so system designers attempt to minimize their use.

Recent advances in fiber and optical communications technology have reduced signal degradation so far that *regeneration* of the optical signal is only needed over distances of hundreds of kilometers. This has greatly reduced the cost of optical networking, particularly over undersea spans where the cost and reliability of repeaters is one of the key factors determining the performance of the whole cable system. The main advances contributing to these performance improvements are dispersion management, which seeks to balance the effects of dispersion against non-linearity; and solitons, which use nonlinear effects in the fiber to enable dispersion-free propagation over long distances.

Last Mile

Although fiber-optic systems excel in high-bandwidth applications, optical fiber has been slow to achieve its goal of fiber to the premises or to solve the last mile problem. However, as bandwidth demand increases, more and more progress towards this goal can be observed. In Japan, for instance EPON has largely replaced DSL as a broadband Internet source. South Korea's KT also provides a service called FTTH (Fiber To The Home), which provides fiber-optic connections to the subscriber's home. The largest FTTH deployments are in Japan, South Korea, and China. Singapore started implementation of their all-fiber Next Generation Nationwide Broadband Network (Next Gen NBN), which is slated for completion in 2012 and is being installed by OpenNet. Since they began rolling out services in September 2010, network coverage in Singapore has reached 85% nationwide.

In the US, Verizon Communications provides a FTTH service called FiOS to select high-ARPU (Average Revenue Per User) markets within its existing territory. The other major surviving ILEC (or Incumbent Local Exchange Carrier), AT&T, uses a FTTN (Fiber To The Node) service called U-verse with twisted-pair to the home. Their MSO competitors employ FTTN with coax using HFC. All of the major access networks use fiber for the bulk of the distance from the service provider's network to the customer.

Also in the US, Wilson Utilities, located in Wilson, North Carolina, has implemented FTTH and has successfully achieved 1 gigabit fiber to the home. This was implemented in late 2013. Wilson Utilities first rolled out their FTTH (Fiber to the Home) in 2012 with speeds offerings of 20/40/60/100 megabits per second. Their service is referred to as GreenLight.

Some other small cities in the US, such as Morristown, TN, have had their local utility company, Morristown Utility Systems in this case, deploy FTTH, offering symmetric gigabit speeds to each subscriber (though most are 50/50 or 100/100 mbit). It's called MUS Fibernet. AT&T and others have aggressively sought legislation at the state level to prevent further competition from municipalities, despite their low investment in rural areas.

The globally dominant access network technology is EPON (Ethernet Passive Optical Network). In Europe, and among telcos in the United States, BPON (ATM-based Broadband PON) and GPON (Gigabit PON) had roots in the FSAN (Full Service Access Network) and ITU-T standards organizations under their control.

Comparison with Electrical Transmission

A mobile fiber optic splice lab used to access and splice underground cables

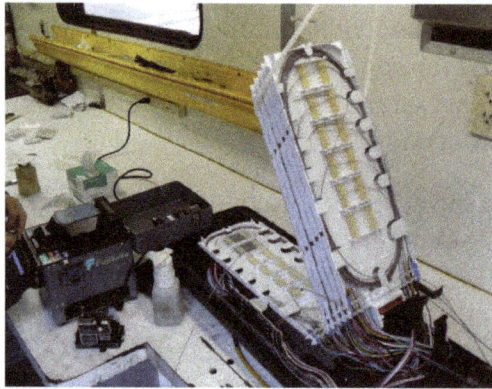

An underground fiber optic splice enclosure opened up

The choice between optical fiber and electrical (or copper) transmission for a particular system is made based on a number of trade-offs. Optical fiber is generally chosen for systems requiring higher bandwidth or spanning longer distances than electrical cabling can accommodate.

The main benefits of fiber are its exceptionally low loss (allowing long distances between amplifiers/repeaters), its absence of ground currents and other parasite signal and power issues common to long parallel electric conductor runs (due to its reliance on light rather than electricity for transmission, and the dielectric nature of fiber optic), and its inherently high data-carrying capacity. Thousands of electrical links would be required to replace a single high bandwidth fiber cable. Another benefit of fibers is that even when run alongside each other for long distances, fiber cables experience effectively no crosstalk, in contrast to some types of electrical transmission lines. Fiber can be installed in areas with high electromagnetic interference (EMI), such as alongside utility lines, power lines, and railroad tracks. Nonmetallic all-dielectric cables are also ideal for areas of high lightning-strike incidence.

For comparison, while single-line, voice-grade copper systems longer than a couple of kilometers require in-line signal repeaters for satisfactory performance; it is not unusual for optical systems to go over 100 kilometers (62 mi), with no active or passive processing. Single-mode fiber cables are commonly available in 12 km lengths, minimizing the number of splices required over a long cable run. Multi-mode fiber is available in lengths up to 4 km, although industrial standards only mandate 2 km unbroken runs.

In short distance and relatively low bandwidth applications, electrical transmission is often preferred because of its

- Lower material cost, where large quantities are not required

- Lower cost of transmitters and receivers

- Capability to carry electrical power as well as signals (in appropriately designed cables)

- Ease of operating transducers in linear mode.

- Crosstalk from nearby cables and other parasitical unwanted signals increase profits from replacement and mitigation devices.

Optical fibers are more difficult and expensive to splice than electrical conductors. And at higher powers, optical fibers are susceptible to fiber fuse, resulting in catastrophic destruction of the fiber core and damage to transmission components.

Because of these benefits of electrical transmission, optical communication is not common in short box-to-box, backplane, or chip-to-chip applications; however, optical systems on those scales have been demonstrated in the laboratory.

In certain situations fiber may be used even for short distance or low bandwidth applications, due to other important features:

- Immunity to electromagnetic interference, including nuclear electromagnetic pulses.

- High electrical resistance, making it safe to use near high-voltage equipment or between areas with different earth potentials.

- Lighter weight—important, for example, in aircraft.

- No sparks—important in flammable or explosive gas environments.

- Not electromagnetically radiating, and difficult to tap without disrupting the signal—important in high-security environments.

- Much smaller cable size—important where pathway is limited, such as networking an existing building, where smaller channels can be drilled and space can be saved in existing cable ducts and trays.

- Resistance to corrosion due to non-metallic transmission medium

Optical fiber cables can be installed in buildings with the same equipment that is used to install copper and coaxial cables, with some modifications due to the small size and limited pull tension and bend radius of optical cables. Optical cables can typically be installed in duct systems in spans of 6000 meters or more depending on the duct's condition, layout of the duct system, and installation technique. Longer cables can be coiled at an intermediate point and pulled farther into the duct system as necessary.

Governing Standards

In order for various manufacturers to be able to develop components that function compatibly in fiber optic communication systems, a number of standards have been developed. The International Telecommunications Union publishes several standards related to the characteristics and performance of fibers themselves, including

- ITU-T G.651, "Characteristics of a 50/125 μm multimode graded index optical fibre cable"

- ITU-T G.652, "Characteristics of a single-mode optical fibre cable"

Other standards specify performance criteria for fiber, transmitters, and receivers to be used together in conforming systems. Some of these standards are:

- 100 Gigabit Ethernet

- 10 Gigabit Ethernet

- Fibre Channel

- Gigabit Ethernet

- HIPPI

- Synchronous Digital Hierarchy

- Synchronous Optical Networking

- Optical Transport Network (OTN)

TOSLINK is the most common format for digital audio cable using plastic optical fiber to connect digital sources to digital receivers.

Fiber Optic Sensor

A fiber optic sensor is a sensor that uses optical fiber either as the sensing element ("intrinsic sensors"), or as a means of relaying signals from a remote sensor to the electronics that process the signals ("extrinsic sensors"). Fibers have many uses in remote sensing. Depending on the application, fiber may be used because of its small size, or because no electrical power is needed at the remote location, or because many sensors can be multiplexed along the length of a fiber by using light wavelength shift for each sensor, or by sensing the time delay as light passes along the fiber through each sensor. Time delay can be determined using a device such as an optical time-domain reflectometer and wavelength shift can be calculated using an instrument implementing optical frequency domain reflectometry.

Fiber optic sensors are also immune to electromagnetic interference, and do not conduct electricity so they can be used in places where there is high voltage electricity or flammable material such as jet fuel. Fiber optic sensors can be designed to withstand high temperatures as well.

Intrinsic Sensors

Optical fibers can be used as sensors to measure strain,temperature, pressure and other quantities by modifying a fiber so that the quantity to be measured modulates the intensity, phase, polarization, wavelength or transit time of light in the fiber. Sensors that vary the intensity of light are the simplest, since only a simple source and detector are required. A particularly useful feature of intrinsic fiber optic sensors is that they can, if required, provide distributed sensing over very large distances.

Temperature can be measured by using a fiber that has evanescent loss that varies with temperature, or by analyzing the Raman scattering of the optical fiber. Electrical voltage can be sensed by nonlinear optical effects in specially-doped fiber, which alter the polarization of light as a function of voltage or electric field. Angle measurement sensors can be based on the Sagnac effect.

Special fibers like long-period fiber grating (LPG) optical fibers can be used for direction recognition . Photonics Research Group of Aston University in UK has some publications on vectorial bend sensor applications.

Optical fibers are used as hydrophones for seismic and sonar applications. Hydrophone systems with more than one hundred sensors per fiber cable have been developed. Hydrophone sensor systems are used by the oil industry as well as a few countries' navies. Both bottom-mounted hydrophone arrays and towed streamer systems are in use. The German company Sennheiser developed a laser microphone for use with optical fibers.

A fiber optic microphone and fiber-optic based headphone are useful in areas with strong electrical or magnetic fields, such as communication amongst the team of people working on a patient inside a magnetic resonance imaging (MRI) machine during MRI-guided surgery.

Optical fiber sensors for temperature and pressure have been developed for downhole measurement in oil wells. The fiber optic sensor is well suited for this environment as it functions at temperatures too high for semiconductor sensors (distributed temperature sensing).

Optical fibers can be made into interferometric sensors such as fiber optic gyroscopes, which are used in the Boeing 767 and in some car models (for navigation purposes). They are also used to make hydrogen sensors.

Fiber-optic sensors have been developed to measure co-located temperature and strain simultaneously with very high accuracy using fiber Bragg gratings. This is particularly useful when acquiring information from small or complex structures. Fiber Bragg grating sensors are also particularly well suited for remote monitoring, and they can be interrogated 250 km away from the monitoring station using an optical fiber cable. Brillouin scattering effects can also be used to detect strain and temperature over large distances (20–120 kilometers).

Other Examples

A fiber-optic AC/DC voltage sensor in the middle and high voltage range (100–2000 V) can be created by inducing measurable amounts of Kerr nonlinearity in single mode optical fiber by exposing a calculated length of fiber to the external electric field. The measurement technique is based on polarimetric detection and high accuracy is achieved in a hostile industrial environment.

High frequency (5 MHz–1 GHz) electromagnetic fields can be detected by induced nonlinear effects in fiber with a suitable structure. The fiber used is designed such that the Faraday and Kerr effects cause considerable phase change in the presence of the external field. With appropriate sensor design, this type of fiber can be used to measure different electrical and magnetic quantities and different internal parameters of fiber material.

Electrical power can be measured in a fiber by using a structured bulk fiber ampere sensor coupled with proper signal processing in a polarimetric detection scheme. Experiments have been carried out in support of the technique.

Fiber-optic sensors are used in electrical switchgear to transmit light from an electrical arc flash to a digital protective relay to enable fast tripping of a breaker to reduce the energy in the arc blast.

Fiber Bragg grating based fiber optic sensors significantly enhance performance, efficiency and safety in several industries. With FBG integrated technology, sensors can provide detailed analysis and comprehensive reports on insights with very high resolution. These type of sensors are used extensively in several industries like telecommunication, automotive, aerospace, energy, etc.Fiber Bragg gratings are sensitive to the static pressure, mechanical tension and compression and fiber temperature changes.

The efficiency of fiber Bragg grating based fiber optic sensors can be provided by means of central wavelength adjustment of light emitting source in accordance with the current Bragg gratings reflection spectra.

Extrinsic Sensors

Extrinsic fiber optic sensors use an optical fiber cable, normally a multimode one, to transmit modulated light from either a non-fiber optical sensor, or an electronic sensor connected to an optical transmitter. A major benefit of extrinsic sensors is their ability to reach places which are otherwise inaccessible. An example is the measurement of temperature inside aircraft jet engines by using a fiber to transmit radiation into a radiation pyrometer located outside the engine. Extrinsic sensors can also be used in the same way to measure the internal temperature of electrical transformers, where the extreme electromagnetic fields present make other measurement techniques impossible.

Extrinsic fiber optic sensors provide excellent protection of measurement signals against noise corruption. Unfortunately, many conventional sensors produce electrical output which must be converted into an optical signal for use with fiber. For example, in the case of a platinum resistance thermometer, the temperature changes are translated into resistance changes. The PRT must therefore have an electrical power supply. The modulated voltage level at the output of the PRT can then be injected into the optical fiber via the usual type of transmitter. This complicates the measurement process and means that low-voltage power cables must be routed to the transducer.

Extrinsic sensors are used to measure vibration, rotation, displacement, velocity, acceleration, torque, and temperature.

Free-space Optical Communication

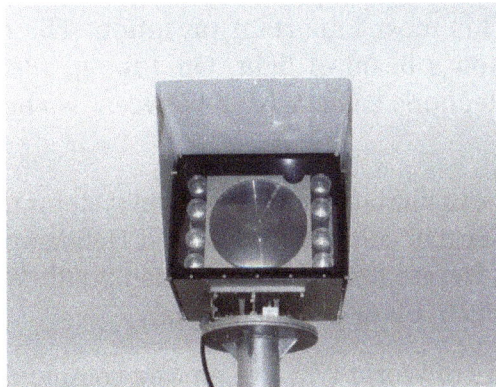

An 8-beam free space optics laser link, rated for 1 Gbit/s. The receptor is the large disc in the middle, the transmitters the smaller ones. At the top right corner is a monocular forassisting the alignment of the two heads.

Free-space optical communication (FSO) is an optical communication technology that uses light propagating in free space to wirelessly transmit data for telecommunications or computer networking. "Free space" means air, outer space, vacuum, or something similar. This contrasts with using solids such as optical fiber cable or an optical transmission line. The technology is useful where the physical connections are impractical due to high costs or other considerations.

History

A photophone receiver and headset, one half of Bell and Tainter's optical telecommunication system of 1880

Optical communications, in various forms, have been used for thousands of years. The Ancient Greeks used a coded alphabetic system of signalling with torches developed by Cleoxenus, Democleitus and Polybius. In the modern era, semaphores and wireless solar telegraphs called heliographs were developed, using coded signals to communicate with their recipients.

In 1880 Alexander Graham Bell and his assistant Charles Sumner Tainter created the Photophone, at Bell's newly established Volta Laboratory in Washington, DC. Bell considered it his most important invention. The device allowed for the transmission of sound on a beam of light. On June 3, 1880, Bell conducted the world's first wireless telephone transmission between two buildings, some 213 meters (700 feet) apart.

Its first practical use came in military communication systems many decades later, first for optical telegraphy. German colonial troops used Heliograph telegraphy transmitters during the 1904/05 Herero Genocide in German South-West Africa (today's Namibia) as did British, French, US or Ottoman signals.

During the trench warfare of World War I when wire communications were often cut, German signals used three types of optical Morse transmitters called *Blinkgerät*, the intermediate type for distances of up to 4 km (2.5 miles) at daylight and of up to 8 km (5 miles) at night, using red filters for undetected communications. Optical telephone communications were tested at the end of the war, but not introduced at troop level. In addition, special blinkgeräts were used for communication with airplanes, balloons, and tanks, with varying success.

WW I German Blinkgerät

A major technological step was to replace the Morse code by modulating optical waves in speech transmission. Carl Zeiss Jena developed the *Lichtsprechgerät 80/80* (literal translation: optical speaking device) that the German army used in their World War II anti-aircraft defense units, or in bunkers at the Atlantic Wall.

The invention of lasers in the 1960s revolutionized free space optics. Military organizations were particularly interested and boosted their development. However the technology lost market momentum when the installation of optical fiber networks for civilian uses was at its peak.

Many simple and inexpensive consumer remote controls use low-speed communication using infrared (IR) light. This is known as consumer IR technologies.

A recently declassified 1987 Pentagon report reveals free-space lasers have been mounted on Israeli F-15 fighter jets for the purposes of surveillance, missile-tracking, and targeted weaponry.

Usage and Technologies

Free-space point-to-point optical links can be implemented using infrared laser light, although low-data-rate communication over short distances is possible using LEDs. Infrared Data Association (IrDA) technology is a very simple form of free-space optical communications. On the communications side the FSO technology is considered as a part of the Optical Wireless Communications applications. Free-space optics can be used for communications between spacecraft.

Current Market Demands

The demand for a high-speed (10 GBps+) and long range (3 – 5 km) FSO system is apparent in the market place.

- In 2008, MRV Communications introduced a free-space optics (FSO)-based system with a data rate of 10GB/s initially claiming a distance of 2 km at high availability. This equipment is no longer available; before end-of-life, the product's useful distance was changed down to 350m.

- In 2013, the company MOSTCOM started to serially produce a new wireless communication system that also had a data rate of 10Gb/s as well as an improved range of up to 2.5 km, but to get to 99.99% up-time the designers used an RF hybrid solution, meaning the data rate drops to extremely low levels during atmospheric disturbances (typically down to 10MB/s).

- LightPointe offers many similar hybrid solutions to MOSTCOM's offering.

Useful Distances

The reliability of FSO units has always been a problem for commercial telecommunications. Consistently, studies find too many dropped packets and signal errors over small ranges (400 to 500 meters). This is from both independent studies, such as in the Czech republic, as well as formal internal nationwide studies, such as one conducted by MRV FSO staff. Military based studies consistently produce longer estimates for reliability, projecting the maximum range for terrestrial links is of the order of 2 to 3 km (1.2 to 1.9 mi). All studies agree the stability and quality of the link is highly dependent on atmospheric factors such as rain, fog, dust and heat.

Extending the Useful Distance

The main reason terrestrial communications have been limited to non-commercial telecommunications functions is fog. Fog consistently keeps FSO laser links over 500 meters from achieving a year-round bit error rate of 99.999%. Several entities are continually attempting to overcome these key disadvantages to FSO communications and field a system with a better quality of service. DARPA has sponsored over $130 million USD in research towards this effort, with the ORCA and ORCLE programs.

Other non-government groups are fielding tests to evaluate different technologies that some claim have the ability to address key FSO adoption challenges. As of October 2014, none have fielded a working system that addresses the most common atmospheric events.

DARPA ORCA Official Concept Art created circa 2008

FSO research from 1998-2006 in the private sector totaled $407.1 million, divided primarily among 4 start-up companies. All four failed to deliver products that would meet telecommunications quality and distance standards:

- Terabeam received approximately $226 million in funding. AT&T and Lucent backed this attempt. The work ultimately failed, and the company reorganized in 2004.

- AirFiber received $96.1 million in funding, and never solved the weather issue. They sold out to MRV communications in 2003, and MRV sold their FSO units until 2012 when the end-of-life was abruptly announced for the Terescope series.

- LightPointe Communications received $76 million in start-up funds, and eventually reorganized to sell hybrid FSO-RF units to overcome the weather-based challenges.

- The Maxima Corporation published its operating theory in Science (magazine), and received $9 million in funding before permanently shutting down. No known spin-off or purchase followed this effort.

One private company published a paper on Nov 20,2014, claiming they had achieved commercial reliability (99.999% availability) in extreme fog. There is no indication this product is currently commercially available.

Extraterrestrial

The massive advantages of laser communication in space have multiple space agencies racing to develop a stable space communication platform, with many significant demonstrations and achievements. To date (18 December 2014), *no laser communication system is in use in space.*

Demonstrations in Space:

The first gigabit laser-based communication was achieved by the European Space Agency and called the European Data Relay System (EDRS) on November 28, 2014. The initial images have just been demonstrated, and a working system is expected to be in place in the 2015-2016 time frame.

NASA's OPALS announced a breakthrough in space-to-ground communication December 9, 2014, uploading 175 megabytes in 3.5 seconds. Their system is also able to re-acquire tracking after the signal was lost due to cloud cover.

In January 2013, NASA used lasers to beam an image of the Mona Lisa to the Lunar Reconnaissance Orbiter roughly 390,000 km (240,000 mi) away. To compensate for atmospheric interference, an error correction code algorithm similar to that used in CDs was implemented.

A two-way distance record for communication was set by the Mercury laser altimeter instrument aboard the MESSENGER spacecraft, and was able to communicate across a distance of 24 million km (15 million miles), as the craft neared Earth on a fly-by in May, 2005. The previous record had been set with a one-way detection of laser light from Earth, by the Galileo probe, of 6 million km in 1992. Quote from Laser Communication in Space Demonstrations (EDRS)

LEDs

RONJA is a free implementation of FSO using high-intensity LEDs.

In 2001, Twibright Labs released Ronja Metropolis, an open source DIY 10Mbit/s full duplex LED FSO over 1.4 km In 2004, a Visible Light Communication Consortium was formed in Japan. This was based on work from researchers that used a white LED-based space lighting system for indoor local area network (LAN) communications. These systems present advantages over traditional UHF RF-based systems from improved isolation between systems, the size and cost of receivers/transmitters, RF licensing laws and by combining space lighting and communication into the same system. In January 2009 a task force for visible light communication was formed by the Institute of Electrical and Electronics Engineers working group for wireless personal area network standards known as IEEE 802.15.7. A trial was announced in 2010 in St. Cloud, Minnesota.

Amateur radio operators have achieved significantly farther distances using incoherent sources of light from high-intensity LEDs. One reported 173 miles (278 km) in 2007. However, physical limitations of the equipment used limited bandwidths to about 4 kHz. The high sensitivities required of the detector to cover such distances made the internal capacitance of the photodiode used a dominant factor in the high-impedance amplifier which followed it, thus naturally forming a low-pass filter with a cut-off frequency in the 4 kHz range. From the other side use of lasers radiation source allows to reach very high data rates which are comparable to fiber communications.

Projected data rates and future data rate claims vary. A low-cost white LED (GaN-phosphor) which could be used for space lighting can typically be modulated up to 20 MHz. Data rates of over 100 Mbit/s can be easily achieved using efficient modulation schemes and Siemens claimed to have achieved over 500 Mbit/s in 2010. Research published in 2009 used a similar system for traffic control of automated vehicles with LED traffic lights.

In September 2013, pureLiFi, the Edinburgh start-up working on Li-Fi, also demonstrated high speed point-to-point connectivity using any off-the-shelf LED light bulb. In previous work, high bandwidth specialist LEDs have been used to achieve the high data rates. The new system, the Li-1st, maximizes the available optical bandwidth for any LED device, thereby reducing the cost and improving the performance of deploying indoor FSO systems.

Engineering Details

Typically, best use scenarios for this technology are:

- LAN-to-LAN connections on campuses at Fast Ethernet or Gigabit Ethernet speeds
- LAN-to-LAN connections in a city, a metropolitan area network
- To cross a public road or other barriers which the sender and receiver do not own
- Speedy service delivery of high-bandwidth access to optical fiber networks
- Converged Voice-Data-Connection
- Temporary network installation (for events or other purposes)

- Reestablish high-speed connection quickly (disaster recovery)

- As an alternative or upgrade add-on to existing wireless technologies

  - Especially powerful in combination with auto aiming systems, this way you could power moving cars or you can power your laptop while you move or use auto-aiming nodes to create a network with other nodes.

- As a safety add-on for important fiber connections (redundancy)

- For communications between spacecraft, including elements of a satellite constellation

- For inter- and intra -chip communication.

The light beam can be very narrow, which makes FSO hard to intercept, improving security. In any case, it is comparatively easy to encrypt any data traveling across the FSO connection for additional security. FSO provides vastly improved electromagnetic interference (EMI) behavior compared to using microwaves.

Technical Advantages

- Ease of deployment

- Can be used to power devices

- License-free long-range operation (in contrast with radio communication)

- High bit rates

- Low bit error rates

- Immunity to electromagnetic interference

- Full duplex operation

- Protocol transparency

- Increased security when working with narrow beam(s)

- No Fresnel zone necessary

- Reference open source implementation

Range Limiting Factors

For terrestrial applications, the principal limiting factors are:

- Fog (10..~100 dB/km attenuation)

- Beam dispersion

- Atmospheric absorption

- Rain

- Snow

- Terrestrial scintillation

- Interference from background light sources (including the Sun)

- Shadowing

- Pointing stability in wind

- Pollution / smog

These factors cause an attenuated receiver signal and lead to higher bit error ratio (BER). To overcome these issues, vendors found some solutions, like multi-beam or multi-path architectures, which use more than one sender and more than one receiver. Some state-of-the-art devices also have larger fade margin (extra power, reserved for rain, smog, fog). To keep an eye-safe environment, good FSO systems have a limited laser power density and support laser classes 1 or 1M. Atmospheric and fog attenuation, which are exponential in nature, limit practical range of FSO devices to several kilometres.

Optical Wireless Communications

Optical wireless communications (OWC) is a form of optical communication in which unguided visible, infrared (IR), or ultraviolet (UV) light is used to carry a signal.

OWC systems operating in the visible band (390–750 nm) are commonly referred to as visible light communication (VLC). VLC systems take advantage of light emitting diodes (LEDs) which can be pulsed at very high speeds without noticeable effect on the lighting output and human eye. VLC can be possibly used in a wide range of applications including wireless local area networks, wireless personal area networks and vehicular networks among others. On the other hand, terrestrial point-to-point OWC systems, also known as the free space optical (FSO) systems, operate at the near IR frequencies (750–1600 nm). These systems typically use laser transmitters and offer a cost-effective protocol-transparent link with high data rates, i.e., 10 Gbit/s per wavelength, and provide a potential solution for the backhaul bottleneck. There has also been a growing interest on ultraviolet communication (UVC) as a result of recent progress in solid state optical sources/detectors operating within solar-blind UV spectrum (200–280 nm). In this so-called deep UV band, solar radiation is negligible at the ground level and this makes possible the design of photon-counting detectors with wide field-of-view receivers that increase the received energy with little additional background noise. Such designs are particularly useful for outdoor non-line-of-sight configurations to support low power short-range UVC such as in wireless sensor and ad-hoc networks.

History

The proliferation of wireless communications stands out as one of the most significant phenomena in the history of technology. Wireless technologies have become essential much more quickly during the last four decades and they will be a key element of society progress for the foreseeable future. The radio-frequency (RF) technologies wide-scale deployment has become the key factor to the wireless devices and systems expansion. However, the electromagnetic spectrum where the wireless systems are deployed is limited in capacity and costly according to its exclusive licenses of exploitation. With the raise of data heavy wireless communications, the demand for RF spectrum is outstripping supply and they become to consider other viable options for wireless communication using the upper parts of the electromagnetic spectrum not just RF.

Optical wireless communication (OWC) refers to transmission in unguided propagation media through the use of optical carriers, i.e., visible, infrared (IR), and ultraviolet (UV) band. Signalling through beacon fires, smoke, ship flags and semaphore telegraph can be considered the historical forms of OWC. Sunlight has been also used for long distance signalling since very early times. The earliest use of sunlight for communication purposes is attributed to ancient Greeks and Romans who used their polished shields to send signals by reflecting sunlight during battles. In 1810, Carl Friedrich Gauss invented the heliograph which involves a pair of mirrors to direct a controlled beam of sunlight to a distant station. Although the original heliograph was designed for geodetic survey, it was used extensively for military purposes during the late 19th and early 20th century. In 1880, Alexander Graham Bell invented the photophone, known as the world's first wireless telephone system.

The military interest on photophone however continued. For example, in 1935, the German Army developed a photophone where a tungsten filament lamp with an IR transmitting filter was used as a light source. Also, American and German military laboratories continued the development of high pressure arc lamps for optical communication until the 1950s. In modern sense, OWC uses either lasers or light emitting diodes (LEDs) as transmitters. In 1962, MIT Lincoln Labs built an experimental OWC link using a light emitting GaAs diode and was able to transmit TV signals over a distance of 30 miles. After the invention of laser, OWC was envisioned to be the main deployment area for lasers and many trials were conducted using different types of lasers and modulation schemes. However, the results were in general disappointing due to large divergence of laser beams and the inability to cope with atmospheric effects. With the development of low-loss fiber optics in the 1970s, they became the obvious choice for long distance optical transmission and shifted the focus away from OWC systems.

Current Status

Over the decades, the interest in OWC remained mainly limited to covert military applications, and space applications including inter-satellite and deep-space links. OWC's mass market penetration has been so far limited with the exception of IrDA which became a highly successful wireless short-range transmission solution. Development of novel and efficient wireless technologies for a range of transmission links is essential for building future heterogeneous communication networks to support a wide range of service types with various traffic patterns and to meet the ever-increasing demands for higher data rates. Variations of OWC can be potentially employed in a diverse range of communication applications ranging from optical interconnects within integrated circuits through outdoor inter-building links to satellite communications.

Photo: ESA; Example of possible ultra-long OWC on inter-satellite European Data Relay Satellite (EDRS) system

OWC long range inter-buildings communications idea for Istanbul Skyline

Applications

Based on the transmission range, OWC can be studied in five categories:

1. Ultra-short range OWC: chip-to-chip communications in stacked and closely packed multi-chip packages.

2. Short range OWC: wireless body area network (WBAN) and wireless personal area network (WPAN) applications under standard IEEE 802.15.7, underwater communications.

3. Medium range OWC: indoor IR and visible light communications (VLC) for wireless local area networks (WLANs) and inter-vehicular and vehicle-to-infrastructure communications.

4. Long range OWC,: inter-building connections, also called Free-Space Optical Communications (FSO).

5. Ultra-long range OWC: inter-satellite links.

Recent Trends

- In January 2015, IEEE 802.15 formed a Task Group to write a revision to IEEE 802.15.7-2011 that accommodates infrared and near ultraviolet wavelengths, in addition to visible light, and adds options such as Optical Camera Communications and LiFi.

- At long range OWC applications a 1 Gbit/s - 60 km range link between ground to aircraft at 800 km/h speed has been demonstrated, "Extreme Test for the ViaLight Laser Communication Terminal MLT-20 – Optical Downlink from a Jet Aircraft at 800 km/h", DLR and EADS December 2013.

- On consumer devices and short-range OWC applications on phones; Charge and receive data with light at your smartphone: TCL Communication/ALCATEL ONETOUCH and Sunpartner Technologies announces the first fully integrated solar smartphone. March 2014.

- On ultra-long range OWC applications the NASA's Lunar Laser Communication Demonstration (LLCD) transmitted data from lunar orbit to Earth at a rate of 622 Megabits-per-second (Mbps), November 2013.

- The Next Generation of OWC / Visible Light Communications demonstrated 10 Mb/s transmission with Polymer Light-Emitting Diodes or OLED.

- On OWC research activities there is a European research project action IC1101 OPTICWISE of the COST Programme (European Cooperation in Science and Technology) funded by the European Science Foundation, allowing the coordination of nationally funded research on a European level. The Action aims to serve as a high-profile consolidated European scientific platform for interdisciplinary optical wireless communication (OWC) research activities. It was launched in November 2011 and will run until November 2015. More than 20 countries represented.

- The consumer and industry OWC technologies adoption is represented by the Li-Fi Consortium, founded in 2011 is a Non-profit organization, devoted to introduce optical wireless technology. Promotes the adoption of Light Fidelity (Li-Fi) products.

- An example of Asian awareness about OWC is the VLCC visible light communication consortium in Japan, established at 2007 in order to realize safe, ubiquitous telecommunication system using visible light through the activities of market research, promotion, and standardization.

- In the USA there are several OWC initiatives, including the "Smart Lighting Engineering Research Center", founded in 2008 by the National Science Foundation (NSF) is a partnership of Rensselaer Polytechnic Institute (lead institution), Boston University and the University of New Mexico. Outreach partners are Howard University, Morgan State University, and Rose-Hulman Institute of Technology.

Li-Fi

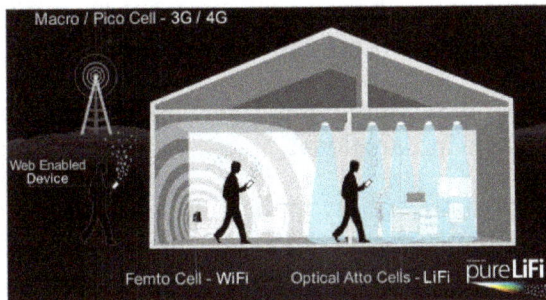

LiFi works in complement with existing and emerging wireless systems.

Light Fidelity (Li-Fi) is a bidirectional, high-speed and fully networked wireless communication technology similar to Wi-Fi. The term was coined by Harald Haas and is a form of visible light communication and a subset of optical wireless communications (OWC) and could be a complement to RF communication (Wi-Fi or cellular networks), or even a replacement in contexts of data broadcasting.

It is wire and uv visible-light communication or infrared and near-ultraviolet instead of radio-frequency spectrum, part of optical wireless communications technology, which carries much more information, and has been proposed as a solution to the RF-bandwidth limitations.

Technology Details

This OWC technology uses light from light-emitting diodes (LEDs) as a medium to deliver networked, mobile, high-speed communication in a similar manner to Wi-Fi. The Li-Fi market is projected to have a compound annual growth rate of 82% from 2013 to 2018 and to be worth over $6 billion per year by 2018.

Visible light communications (VLC) works by switching the current to the LEDs off and on at a very high rate, too quick to be noticed by the human eye. Although Li-Fi LEDs would have to be kept on to transmit data, they could be dimmed to below human visibility while still emitting enough light to carry data. The light waves cannot penetrate walls which makes a much shorter range, though more secure from hacking, relative to Wi-Fi. Direct line of sight is not necessary for Li-Fi to transmit a signal; light reflected off the walls can achieve 70 Mbit/s.

Li-Fi has the advantage of being useful in electromagnetic sensitive areas such as in aircraft cabins, hospitals and nuclear power plants without causing electromagnetic interference. Both Wi-Fi and Li-Fi transmit data over the electromagnetic spectrum, but whereas Wi-Fi utilizes radio waves, Li-Fi uses visible light. While the US Federal Communications Commission has warned of a potential spectrum crisis because Wi-Fi is close to full capacity, Li-Fi has almost no limitations on capacity. The visible light spectrum is 10,000 times larger than the entire radio frequency spectrum. Researchers have reached data rates of over 10 Gbit/s, which is much faster than typical fast broadband in 2013. Li-Fi is expected to be ten times cheaper than Wi-Fi. Short range, low reliability and high installation costs are the potential downsides.

PureLiFi demonstrated the first commercially available Li-Fi system, the Li-1st, at the 2014 Mobile World Congress in Barcelona.

Bg-Fi is a Li-Fi system consisting of an application for a mobile device, and a simple consumer product, like an IoT (Internet of Things) device, with color sensor, microcontroller, and embedded software. Light from the mobile device display communicates to the color sensor on the consumer product, which converts the light into digital information. Light emitting diodes enable the consumer product to communicate synchronously with the mobile device.

History

Harald Haas, coined the term "Li-Fi" at his TED Global Talk where he introduced the idea of "Wireless data from every light". He is Chairman of Mobile Communications at the University of Edinburgh and co-founder of pureLiFi.

The general term visible light communication (VLC), whose history dates back to the 1880s, includes any use of the visible light portion of the electromagnetic spectrum to transmit information. The D-Light project at Edinburgh's Institute for Digital Communications was funded from January 2010 to January 2012. Haas promoted this technology in his 2011 TED Global talk and helped start a company to market it. PureLiFi, formerly pureVLC, is an original equipment manufacturer (OEM) firm set up to commercialize Li-Fi products for integration with existing LED-lighting systems.

In October 2011, companies and industry groups formed the Li-Fi Consortium, to promote high-speed optical wireless systems and to overcome the limited amount of radio-based wireless spectrum available by exploiting a completely different part of the electromagnetic spectrum.

A number of companies offer uni-directional VLC products, which is not the same as Li-Fi - a term defined by the IEEE 802.15.7r1 standardization committee.

VLC technology was exhibited in 2012 using Li-Fi. By August 2013, data rates of over 1.6 Gbit/s were demonstrated over a single color LED. In September 2013, a press release said that Li-Fi, or VLC systems in general, do not require line-of-sight conditions. In October 2013, it was reported Chinese manufacturers were working on Li-Fi development kits.

In April 2014, the Russian company Stins Coman announced the development of a Li-Fi wireless local network called BeamCaster. Their current module transfers data at 1.25 gigabytes per second but they foresee boosting speeds up to 5 GB/second in the near future. In 2014 a new record was established by Sisoft (a Mexican company) that was able to transfer data at speeds of up to 10 Gbit/s across a light spectrum emitted by LED lamps.

Standards

Like Wi-Fi, Li-Fi is wireless and uses similar 802.11 protocols; but it uses visible light communication (instead of radio frequency waves), which has much wider bandwidth.

One part of VLC is modeled after communication protocols established by the IEEE 802 workgroup. However, the IEEE 802.15.7 standard is out-of-date, it fails to consider the latest technological developments in the field of optical wireless communications, specifically with the introduction of optical orthogonal frequency-division multiplexing (O-OFDM) modulation methods which have been optimized for data rates, multiple-access and energy efficiency. The introduction of O-OFDM means that a new drive for standardization of optical wireless communications is required.

Nonetheless, the IEEE 802.15.7 standard defines the physical layer (PHY) and media access control (MAC) layer. The standard is able to deliver enough data rates to transmit audio, video and multimedia services. It takes into account optical transmission mobility, its compatibility with artificial lighting present in infrastructures, and the interference which may be generated by ambient lighting. The MAC layer permits using the link with the other layers as with the TCP/IP protocol.

The standard defines three PHY layers with different rates:

- The PHY I was established for outdoor application and works from 11.67 kbit/s to 267.6 kbit/s.

- The PHY II layer permits reaching data rates from 1.25 Mbit/s to 96 Mbit/s.

- The PHY III is used for many emissions sources with a particular modulation method called color shift keying (CSK). PHY III can deliver rates from 12 Mbit/s to 96 Mbit/s.

The modulation formats recognized for PHY I and PHY II are on-off keying (OOK) and variable pulse position modulation (VPPM). The Manchester coding used for the PHY I and PHY II layers includes the clock inside the transmitted data by representing a logic 0 with an OOK symbol "01" and a logic 1 with an OOK symbol "10", all with a DC component. The DC component avoids light extinction in case of an extended run of logic 0's.

The first VLC smartphone prototype was presented at the Consumer Electronics Show in Las Vegas from January 7–10 in 2014. The phone uses SunPartner's Wysips CON-NECT, a technique that converts light waves into usable energy, making the phone capable of receiving and decoding signals without drawing on its battery. A clear thin layer of crystal glass can be added to small screens like watches and smartphones that make them solar powered. Smartphones could gain 15% more battery life during a typical day. This first smartphones using this technology should arrive in 2015. This screen can also receive VLC signals as well as the smartphone camera. The cost of these screens per smartphone is between $2 and $3, much cheaper than most new technology.

Philips lighting company has developed a VLC system for shoppers at stores. They have to download an app on their smartphone and then their smartphone works with the LEDs in the store. The LEDs can pinpoint where they are located in the store and give them corresponding coupons and information based on which aisle they are on and what they are looking at.

References

- Semaphore Signalling ISBN 978-0-86341-327-8 Communications: an international history of the formative years R. W. Burns, 2004

- Nishizawa, Jun-ichi & Suto, Ken (2004). "Terahertz wave generation and light amplification using Raman effect". In Bhat, K. N. & DasGupta, Amitava. Physics of semiconductor devices. New Delhi, India: Narosa Publishing House. p. 27. ISBN 81-7319-567-6.

- Hecht, Jeff (1999). City of Light, The Story of Fiber Optics. New York: Oxford University Press. p. 114. ISBN 0-19-510818-3.

- Bănică, Florinel-Gabriel (2012). Chemical Sensors and Biosensors: Fundamentals and Applications. Chichester: John Wiley and Sons. Ch. 18–20. ISBN 978-0-470-71066-1.

- Williams, E. A. (ed.) (2007). National Association of Broadcasters Engineering Handbook (Tenth Edition). Taylor & Francis. pp. 1667–1685 (Chapter 6.10 – Fiber–Optic Transmission Systems, Jim Jachetta). ISBN 978-0-240-80751-5.

- Gowar, John (1993). Optical communication systems (2d ed.). Hempstead, UK: Prentice-Hall. p. 209. ISBN 0-13-638727-6.

- "Future Trends in Fiber Optics Communication" (PDF). WCE, London UK. July 2, 2014. ISBN 978-988-19252-7-5.

- Mary Kay Carson (2007). Alexander Graham Bell: Giving Voice To The World. Sterling Biographies. New York: Sterling Publishing. pp. 76–78. ISBN 978-1-4027-3230-0.

- Measures, Raymond M. (2001). Structural Monitoring with Fiber Optic Technology. San Diego, California, USA: Academic Press. pp. Chapter 7. ISBN 0-12-487430-4.

- Mary Kay Carson (2007). Alexander Graham Bell: Giving Voice To The World. Sterling Biographies. New York: Sterling Publishing. pp. 76–78. ISBN 978-1-4027-3230-0.

- pureVLC (10 September 2013). "pureVLC Demonstrates Li-Fi Using Reflected Light". Edinburgh. Retrieved 17 June 2016.

- Vega, Anna (14 July 2014). "Li-fi record data transmission of 10Gbps set using LED lights". Engineering and Technology Magazine. Retrieved 29 November 2015.

Permissions

We would like to thank the editorial team for lending their expertise to make the book truly unique. They have played a crucial role in the development of this book. Without their invaluable contributions this book wouldn't have been possible. They have made vital efforts to compile up to date information on the varied aspects of this subject to make this book a valuable addition to the collection of many professionals and students.

This book was conceptualized with the vision of imparting up-to-date and integrated information in this field. To ensure the same, a matchless editorial board was set up. Every individual on the board went through rigorous rounds of assessment to prove their worth. After which they invested a large part of their time researching and compiling the most relevant data for our readers.

The editorial board has been involved in producing this book since its inception. They have spent rigorous hours researching and exploring the diverse topics which have resulted in the successful publishing of this book. They have passed on their knowledge of decades through this book. To expedite this challenging task, the publisher supported the team at every step. A small team of assistant editors was also appointed to further simplify the editing procedure and attain best results for the readers.

Apart from the editorial board, the designing team has also invested a significant amount of their time in understanding the subject and creating the most relevant covers. They scrutinized every image to scout for the most suitable representation of the subject and create an appropriate cover for the book.

The publishing team has been an ardent support to the editorial, designing and production team. Their endless efforts to recruit the best for this project, has resulted in the accomplishment of this book. They are a veteran in the field of academics and their pool of knowledge is as vast as their experience in printing. Their expertise and guidance has proved useful at every step. Their uncompromising quality standards have made this book an exceptional effort. Their encouragement from time to time has been an inspiration for everyone.

The publisher and the editorial board hope that this book will prove to be a valuable piece of knowledge for students, practitioners and scholars across the globe.

Index

www.ingramcontent.com/pod-product-compliance
Lightning Source LLC
Chambersburg PA
CBHW061933190326
41458CB00009B/2727